pytest 测试实战

[美] Brian Okken 著

陆阳 饶勇 译
赏味不足 审校

华中科技大学出版社
中国·武汉

内 容 简 介

pytest是动态编程语言Python专用的测试框架,它具有易于上手、功能强大、第三方插件丰富、效率高、可扩展性好、兼容性强等特点。本书深入浅出地讲解了pytest的使用方法,尤其是具有特色的fixture的用法。作者通过丰富的测试实例,手把手教读者编写简洁、易于维护的测试代码和插件,让你轻松掌握这个业界最受欢迎的Python测试工具。

Python Testing with pytest © 2017 The Pragmatic Programmers, LLC. All rights reserved.

湖北省版权局著作权合同登记　图字:17-2018-200 号

图书在版编目(CIP)数据

pytest测试实战 /（美）布赖恩·奥肯(Brian Okken) 著 ;陆阳,饶勇译. —武汉:华中科技大学出版社,2018.8 (2021.8重印)
　ISBN 978-7-5680-4442-4

　Ⅰ.①p… Ⅱ.①布… ②陆… ③饶… Ⅲ.①软件工具-程序设计 Ⅳ.①TP311.561

中国版本图书馆 CIP 数据核字(2018)第 164840 号

pytest 测试实战　　　　　　　　　　　　　　　　　　　　　　　　[美]Brian Okken 著
pytest Ceshi Shizhan　　　　　　　　　　　　　　　　　　　　　　　陆阳　饶勇 译

策划编辑：徐定翔
责任编辑：徐定翔
责任监印：周治超

出版发行：华中科技大学出版社(中国·武汉)　　　电话：(027)81321913
　　　　　武汉市东湖新技术开发区华工科技园　　　邮编：430223
录　　排：华中科技大学惠友文印中心
印　　刷：湖北新华印务有限公司
开　　本：787mm×960mm　1/16
印　　张：15
字　　数：361 千字
版　　次：2021年 8月第 1 版第 3 次印刷
定　　价：69.90 元

本书若有印装质量问题,请向出版社营销中心调换
全国免费服务热线：400-6679-118　竭诚为您服务
版权所有　侵权必究

致谢
Acknowledgments

我首先要感谢妻子 Michelle，她也是我的挚友。我的书房是这样的：稿件都堆放在一张略有年代感的橡木方桌上，玻璃书柜上都是我这些年来收集的太空玩具、技术书籍、电路板、杂耍球。铝制储物箱内堆满了便签和装订线，以及宣传书籍用的火箭贴纸。房间一面墙被我和妻子几年前买的天鹅绒覆盖。墙面这样处理可以在录制音频时吸收回声。我喜欢在这里写作，不仅因为这里符合我的个性，更因为这里是 Michelle 与我共同打造的。她总是站在我这边，每当我在博客、播客上发表一些奇思异想时，她都会支持我，包括我计划编写本书。她帮助我确保了写作时间和空间，每当我感到疲惫不堪时，她便会提醒我休息，正如我们上大学时一起熬夜做功课一样。如果没有 Michelle，我的工作和生活将举步维艰。

我有两个聪慧可爱的女儿，Gabriella 和 Sophia，她们都是我的粉丝。Gabriella 和别人聊起编程时，都要提醒别人听我的播客节目；Sophia 上二年级时，书包上贴的都是 Test & Code 字样的贴纸。

我还要感谢编辑 Katharine Dvorak，她帮助我梳理了凌乱的想法，我才能将这些观点系统化地整理成书籍出版。起初我就像写博客一样，文章中夹杂了许多冗余的标题、符号。在 Katharine Dvorak 的悉心指导下，我成为了一名合格的作者。

致谢

感谢 Susannah Davidson Pfalzer、Andy Hunt，以及 The Pragmatic Bookshelf 出版公司的所有人。

技术审校在 pytest 规范和 Python 编码规范方面帮助我改正了许多错误，所以本书的示例代码才能遵循 PEP 8 规范。谢谢大家：Oliver Bestwalter、Florian Bruhin、Floris Bruynooghe、Mark Goody、Peter Hampton、Dave Hunt、Al Krinker、Lokesh Kumar Makani、Bruno Oliveira、Ronny Pfannschmidt、Raphael Pierzina、Luciano Ramalho、Frank Ruiz、Dmitry Zinoviev。这几位也都是 pytest 的核心开发者，或者热门插件的作者。

我还要感谢审校者 Luciano。我曾经把前四章书稿发给 Luciano 审阅，他的审校反馈意见毫无疑问是最苛刻的。我虽然没有完全采纳他的意见，但是几乎重写了前三章的内容，他的反馈意见也影响了我后续章节的写作思路。

感谢 pytest 团队创造出了 pytest 这样棒的工具。再次感谢 Oliver Bestwalter、Florian Bruhin、Floris Bruynooghe、Dave Hunt、Holger Krekel、Bruno Oliveira、Ronny Pfannschmidt、Raphael Pierzina，以及解答我疑惑的所有人。

最后我要感谢那些曾经感谢过我的人。曾有读者发邮件说我写的东西帮他们简化了工作，节约了时间。这对我来说是莫大的鼓励，感谢你们。

Brian Okken
2017 年 9 月

前言
Preface

Python 语言越来越流行，不单在软件开发领域，它还深入数据分析、科学研究、测量等领域。随着 Python 在这些重要领域的发展，高效可靠地进行软件测试变得尤为重要。此外，越来越多的软件开发项目开始使用持续集成，而自动化测试正是其中的关键一环。由于软件迭代周期不断缩短，所以手工测试已经行不通。项目团队必须确定测试结果是可靠的，才能放心地发布产品。

于是 pytest 派上用场。

pytest 是什么？
What Is pytest?

pytest 是一款强大的 Python 测试工具，可以胜任各种类型或级别的软件测试工作，既适合开发团队、QA 团队、独立的测试小组使用，又适合练习测试驱动开发的个人，以及开源团队使用。实际上，越来越多的互联网项目开始放弃 unittest 和 nose，转而使用 pytest，比如 Mozilla 和 Dropbox。因为 pytest 会提供更丰富的功能，包括 assert 重写、第三方插件，以及其他测试工具无法比拟的 fixture 模型。

pytest 是一个软件测试框架。它是一款命令行工具，可以自动找到测试用例执行,并且汇报测试结果。它有丰富的基础库,可以大幅提高用户编写测试用例的效率。它具备可扩展性,用户可以自己编写插件,或者安装第三方提供的插件。pytest 可以直接测试各类 Python 程序,也可以很容易地与其他工具集成到一起使用,比如持续集成、Web 端自动化测试等。

下面列举了一些 pytest 优于其他测试框架的地方。

- 简单的测试可以很简单地编写;
- 复杂的测试也可以很简单地编写;
- 测试的可读性强;
- 易于上手;
- 断言测试失败仅使用原生 assert 关键字,而不是 self.assertEqual(),或者 self.assertLessThan();
- pytest 可以运行由 unittest 和 nose 编写的测试用例。

pytest 项目是由一个正在快速壮大的社区开发和维护的。它灵活、扩展性好,可以很容易地融入已有的开发测试流程。它不依赖于 Python 版本,Python 2(2.6 及更高版本)和 Python 3(3.3 及更高版本)都可以安装最新版本的 pytest。

通过实用示例学习
Learn pytest While Testing an Example Application

没有人愿意学习实际工作中永远不会出现的例子,我也一样。本书将围绕一个实用的测试项目展开学习,其中的测试技巧很容易运用到你的实际工作中。

Tasks 项目
The Tasks Project

我们的测试对象是一个叫 Tasks 的应用程序。Tasks 是一个小巧的任务进度追踪程序，它有一个命令行交互界面，设计结构与很多应用程序相似。我希望读者在编写针对 Tasks 项目测试用例的同时，能够轻松地掌握相关技巧，并运用到自己的测试项目中去。

虽然 Tasks 程序通过 CLI（command-line interface）交互，但其底层编码是通过调用 API 实现的，因此我们的测试工作主要面向的是 API。API 与数据库控制层交互，数据库控制层与文档数据库（MongoDB 或 TinyDB）交互。数据库的类型是在数据库初始化时配置好的。

在进一步讨论 API 之前，我们先了解一下 Tasks 所使用的命令行交互工具。以下是一段示例代码：

```
$ tasks add 'do something' --owner Brian
$ tasks add 'do something else'
$ tasks list
  ID   owner  done  summary
  --   -----  ----  -------
   1   Brian  False do something
   2          False do something else
$ tasks update 2 --owner Brian
$ tasks list
  ID   owner  done  summary
  --   -----  ----  -------
   1   Brian  False do something
   2   Brian  False do something else
$ tasks update 1 --done True
$ tasks list
  ID   owner  done  summary
  --   -----  ----  -------
   1   Brian  True  do something
   2   Brian  False do something else
$ tasks delete 1
$ tasks list
  ID   owner  done  summary
  --   -----  ----  -------
   2   Brian  True  do something
$
```

这个任务管理程序并不复杂，但作为我们的示例已经够用。

测试方法
Test Strategy

虽然 pytest 可以用于单元测试、集成测试、系统测试（端到端测试）、功能测试，但 Tasks 项目将主要采用皮下测试（subcutaneous test）。以下是一些有关测试方法的基本定义。

单元测试　检查一小块代码（如一个函数，或一个类）的测试。第 1 章中的测试就是针对 Tasks 数据结构的单元测试。

集成测试　检查大段的代码（如多个类，或一个子系统）的测试。集成测试的规模介于单元测试与系统测试之间。

系统测试　检查整个系统的测试，通常要求测试环境尽可能接近最终用户的使用环境。

功能测试　检查单个系统功能的测试。就 Tasks 项目而言，针对增加、删除、更新一个任务条目的测试就属于功能测试。

皮下测试　不针对最终用户界面，而是针对用户界面以下的接口的测试。本书中的大部分测试都是针对 API 的，而不是针对 CLI 的，因此这些测试都属于皮下测试。

本书结构
How This Book Is Organized

第 1 章介绍 pytest 的安装，同时会介绍 Tasks 项目的数据结构部分（名为 Task 的 namedtuple），并用它作为测试示例。我们会学习如何指定测试文件运行，以及 pytest 常用的命令行命令，包括重新运行失败测试、遇到失败即停止所有测试、控制堆栈跟踪、控制日志输出，等等。

第 2 章将使用 pip 在本地安装 Tasks 项目，学习在 Python 项目中如何组织测试目录，这样才能针对实际项目编写测试用例。这一章的所有示例都依赖外部程序，包括数据库写入。第 2 章的重点是测试函数，你将学习在 pytest 中高效使用断言语句。这一章还会讲解 marker 标记功能的用法，marker 标记可以将测试进行归类或分组，方便一起运行，也可以将某些测试标记为 skip（跳过不执行），marker 标记还可以告诉 pytest 我们知道某些测试是一定会失败的。如果希望运行指定的测试子集，除了使用 marker，还可以将测试代码组织成测试目录、测试模块、测试类，然后运行。

并非所有的测试代码都要放到测试函数中。第 3 章介绍如何将测试数据、启动逻辑、销毁逻辑放入 fixture（pytest 定义的一种测试脚手架）。设置系统（或子系统、系统单元）是软件测试的重要环节，第 3 章将介绍用一个简单的 fixture 完成这方面的工作（包括对数据库进行初始化，写入数据以备测试之用）。Fixture 模块的功能非常强大，你可以利用它简化测试代码，从而提高代码的可读性和可维护性。Fixture 像测试函数一样，也有参数。利用参数，你只需要编写一份代码，就可以针对 TinyDB 和 MongoDB（或其他 Tasks 项目支持的数据库）开展测试。

第 4 章介绍 pytest 内置的 fixture 以满足测试中常见的一些需求，包括生成和销毁临时目录、截取输出流（通过日志判定结果）、使用 monkey patch、检查是否发出警告，等等。

第 5 章讲解如何在 pytest 中添加命令行选项，如何改进打印输出，如何打包分发自己编写的插件，如何共享定制化的 pytest（包括 fixture）。这一章开发的插件可以改善 Tasks 项目测试失败时的输出呈现方式。你还将学习测试自己的测试插件(元测试)。读完这一章，想必你已经等不及编写自己的插件了。附录 C 收集了一些热门的社区插件，可供参考。

第 6 章讲解通过 pytest.ini 文件修改默认配置，自定义 pytest 的运行方式。pytest.ini 文件可以存放某些命令选项，从而减少你重复输入命令的次数；利用它还可以指定 pytest 忽略某些测试目录，或者指定 pytest 的最低版本，等等。使用 tox.ini 和 setup.cfg 文件也可以实现同样的功能。

第 7 章（最后一章）介绍 pytest 与其他工具的结合使用。我们将借助 tox 让 Tasks 项目在多个 Python 版本上运行；学习如何测试 Tasks 项目的 CLI 部分，而不必 mock 系统的其余部分；借助 coverage.py 检查 Tasks 项目代码块的测试覆盖情况；通过 Jenkins 发起测试并实时显示结果。最后，还会学习如何让 pytest 运行基于 unittest 的测试用例，以及把 pytest 的 fixture 共享给 unittest 的测试用例使用。

阅读基础
What You Need to Know

Python

你不必精通 Python，本书的示例语法很常见。

pip

本书使用 pip 安装 pytest 及其组件。获取新版本的 pip，请查阅附录 B。

命令行

我自己使用的是 Mac 笔记本和 bash，实际上，我只使用了 cd（进入指定目录）和 pytest 两条命令。Windows 系统和所有 UNIX 派生系统都有 cd 命令，所以本书的示例也应该适用于这些操作系统。

满足以上条件，就可以阅读本书了。不擅长编程的人一样可以学习使用 pytest 进行自动化测试。

示例代码与共享资源
Example Code and Online Resources

本书示例代码是使用 Python 3.6 和 pytest 3.2 编写的。pytest 3.2 支持 Python 2.6、Python 2.7、Python 3.3 及以后的版本。

Tasks 项目的源代码及测试用例可以从 pragprog.com 网站上本书的页面下载[1]。如果只是为了理解书中的测试，不必下载源代码，这些代码都通俗易懂。但如果你希望深入地理解 Tasks 项目，以便将来更好地运用学到的技巧，那就有必要下载源代码。我们在本书的页面上还提供了最新的勘误信息[2]。

我编程已有二十余年，从未遇到过像 pytest 这样优秀的测试工具。希望各位阅读本书后能够有所收获，也喜欢上 pytest。

---

[1] https://pragprog.com/titles/bopytest/source_code
[2] https://pragprog.com/titles/bopytest/errata

目录 Contents

第 1 章　pytest 入门 1
  1.1　资源获取 4
  1.2　运行 Pytest 5
  1.3　运行单个测试用例 10
  1.4　使用命令行选项 10
      --collect-only 选项 11
      -k 选项 11
      -m 选项 12
      -x 选项 13
      --maxfail=num 15
      -s 与 --capture=method 16
      --lf（--last-failed）选项 16
      --ff（--failed-first）选项 17
      -v（--verbose）选项 17
      -q（--quiet）选项 18
      -l（--showlocals）选项 19
      --tb=style 选项 20
      --duration=N 选项 21
      --version 选项 22
      -h（--help）选项 23
  1.5　练习 24
  1.6　预告 25

第 2 章　编写测试函数 27
- 2.1　测试示例程序 27
  - 本地安装 Tasks 项目程序包 30
- 2.2　使用 assert 声明 32
- 2.3　预期异常 35
- 2.4　测试函数的标记 36
  - 完善冒烟测试 38
- 2.5　跳过测试 40
- 2.6　标记预期会失败的测试 43
- 2.7　运行测试子集 45
  - 单个目录 45
  - 单个测试文件/模块 46
  - 单个测试函数 47
  - 单个测试类 47
  - 单个测试类中的测试方法 48
  - 用测试名划分测试集合 48
- 2.8　参数化测试 49
- 2.9　练习 56
- 2.10　预告 57

第 3 章　pytest Fixture 59
- 3.1　通过 conftest.py 共享 fixture 60
- 3.2　使用 fixture 执行配置及销毁逻辑 61
- 3.3　使用--setup-show 回溯 fixture 的执行过程 63
- 3.4　使用 fixture 传递测试数据 64
- 3.5　使用多个 fixture 66
- 3.6　指定 fixture 作用范围 68
  - 修改 Tasks 项目的 fixture 作用范围 70
- 3.7　使用 usefixtures 指定 fixture 73
- 3.8　为常用 fixture 添加 autouse 选项 74
- 3.9　为 fixture 重命名 75
- 3.10　Fixture 的参数化 77
- 3.11　参数化 Tasks 项目中的 fixture 80
- 3.12　练习 83

3.13 预告 83

第 4 章 内置 Fixture 85
4.1 使用 tmpdir 和 tmpdir_factory 86
　　在其他作用范围内使用临时目录 88
4.2 使用 pytestconfig 90
4.3 使用 cache 92
4.4 使用 capsys 100
4.5 使用 monkeypatch 102
4.6 使用 doctest_namespace 106
4.7 使用 recwarn 109
4.8 练习 110
4.9 预告 111

第 5 章 插件 113
5.1 寻找插件 114
5.2 安装插件 114
　　从 PyPI 安装 114
　　从 PyPI 安装指定版本 115
　　从 .tar.gz 或 .whl 文件安装 115
　　从本地目录安装 115
　　从 Git 存储仓库安装 116
5.3 编写自己的插件 116
5.4 创建可安装插件 121
5.5 测试插件 125
5.6 创建发布包 129
　　通过共享目录分发插件 130
　　通过 PyPI 发布插件 130
5.7 练习 131
5.8 预告 131

第 6 章 配置 133
6.1 理解 pytest 的配置文件 133
　　用 pytest --help 查看 ini 文件选项 135
　　插件可以添加 ini 文件选项 135
6.2 更改默认命令行选项 136

6.3	注册标记来防范拼写错误	136
6.4	指定 pytest 的最低版本号	138
6.5	指定 pytest 忽略某些目录	138
6.6	指定测试目录	139
6.7	更改测试搜索的规则	141
6.8	禁用 XPASS	142
6.9	避免文件名冲突	143
6.10	练习	145
6.11	预告	145

第 7 章　pytest 与其他工具的搭配使用 147

7.1	pdb：调试失败的测试用例	147
7.2	coverage.py：判断测试覆盖了多少代码	151
7.3	mock：替换部分系统	155
7.4	tox：测试多种配置	162
7.5	Jenkins CI：让测试自动化	166
7.6	unittest：用 pytest 运行历史遗留测试用例	173
7.7	练习	179
7.8	预告	180

附录 A　虚拟环境 181

附录 B　Pip 183

附录 C　常用插件 187

C.1	改变测试流程的插件	187
	pytest-repeat：重复运行测试	187
	pytest-xdist：并行运行测试	189
	pytest-timeout：为测试设置时间限制	190
C.2	改善输出效果的插件	191
	pytest-instafail：查看错误的详细信息	191
	pytest-sugar：显示色彩和进度条	192
	pytest-emoji：为测试增添一些乐趣	193
	pytest-html：为测试生成 HTML 报告	195
C.3	静态分析用的插件	197
	pytest-pycodestyle 和 pytest-pep8：Python 代码风格检查	197

pytest-flake8：更多的风格检查 197
    C.4 Web 开发用的插件 198
        pytest-selenium：借助浏览器完成自动化测试 198
        pytest-django：测试 Django 应用 198
        pytest-flask：测试 Flask 应用 199

附录 D　打包和发布 Python 项目 201
    D.1 创建可安装的模块 201
    D.2 创建可安装的包 203
    D.3 创建源码发布包和 Wheel 文件 205
    D.4 创建可以从 PyPI 安装的包 209

附录 E　xUnit Fixture 211
    E.1 xUnit Fixture 的语法 211
    E.2 混合使用 pytest Fixture 和 xUnit Fixture 214
    E.3 xUnit Fixture 的限制 215

索引 216

第 1 章

pytest 入门
Getting Started with pytest

这是一个测试用例:

```
ch1/test_one.py
def test_passing():
    assert (1, 2, 3) == (1, 2, 3)
```

打开终端执行以下代码:

```
$ cd /path/to/code/ch1
$ pytest test_one.py
=================test session starts==================
collected 1 items
test_one.py .
================1 passed in 0.01 seconds =============
```

test_one.py 后方的一个点号(.)表示:运行了一个测试用例,且测试通过。如果想查看详情,可以在 pytest 后面加上 -v 或者 --verbose 选项。

```
$ pytest -v test_one.py
================test session starts ==================
collected 1 items
test_one.py::test_passing PASSED
================1 passed in 0.01 seconds =============
```

如果你使用的是彩色终端,那么 PASSED 和底部线条是绿色的。以下是一

个注定要失败的测试用例:

`ch1/test_two.py`
```python
def test_failing():
    assert (1, 2, 3) == (3, 2, 1)
```

pytest 展示的失败信息非常清楚,这是它受欢迎的原因之一。

```
$ pytest test_two.py
===============test session starts =================
collected 1 items
test_two.py F
============== FAILURES ===================
_____test_failing _____
def test_failing():
>     assert (1, 2, 3) == (3, 2, 1)
E     assert (1, 2, 3) == (3, 2, 1)
E       At index 0 diff: 1 != 3
E       Use -v to get the full diff
test_two.py:2: AssertionError
============ 1 failed in 0.04 seconds ==============
```

pytest 有一块专门的区域展示 test_failing 的失败信息,它能准确地指出失败原因:index 0 is mismatch。重要的提示信息用红色字体显示(在彩色显示器上),以方便用户阅读。其中有一条提示指出,使用 -v 可以得到更完整的前后对比信息,下面来试一试。

```
$ pytest -v test_two.py
===================== test session starts =====================
collected 1 items
test_two.py::test_failing FAILED
=========================== FAILURES ===========================
_____ test_failing _____
    def test_failing():
>       assert (1, 2, 3) == (3, 2, 1)
E       assert (1, 2, 3) == (3, 2, 1)
E         At index 0 diff: 1 != 3
E         Full diff:
E         - (1, 2, 3)
E         ?    ^    ^
E         + (3, 2, 1)
E         ?    ^    ^
test_two.py:2: AssertionError
==================== 1 failed in 0.04 seconds ====================
```

pytest 添加了几个脱字符（^），准确地指出了前后的区别。pytest 除了易读、易写、易运行、失败提示信息清晰，还有许多优点，如果你有耐心，请听我慢慢告诉你。在我心目中，pytest 是最优秀的测试框架。

本章讲解 pytest 的安装、pytest 的各种运行方式，以及最常用的命令行选项。后面几章的学习内容包括：如何最大限度地发挥 pytest 的优势来编写测试函数；如何编写 fixture（用于存放相同的启动逻辑和销毁逻辑）；如何使用 fixture 和 pytest 插件简化测试工作。

抱歉，我在这里使用了 assert (1, 2, 3) == (3, 2, 1)这样无聊的例子，实际软件测试中显然不会出现这样的例子。接下来，我会以一个名叫 Tasks 的软件项目为示例来编写测试。这个项目并不复杂，很容易理解，同时也比较接近真实的软件项目。

软件测试的一个目标是验证你的猜想——猜想软件的内部逻辑，包括第三方的模块、代码包，甚至 Python 内建的数据结构是如何运作的。Tasks 项目使用名为 Task 的数据结构，它是用 namedtuple 工厂函数生成的。namedtuple 是 Python 标准库的一部分。Task 用于在 UI 层和 API 层之间传递信息，我用它来演示 pytest 的运行，以及一些常用命令行选项的用法。

以下是一个 Task 结构：

```python
from collections import namedtuple
Task = namedtuple('Task', ['summary', 'owner', 'done', 'id'])
```

namedtuple()工厂方法在 Python 2.6 中就已经出现，但我发现许多 Python 开发者并不了解它。我觉得使用 Task 结构比使用(1, 2, 3) == (1, 2, 3)或者 add(1, 2) == 3 有趣得多。

在进一步讲解之前，我们先看看 pytest 的下载和安装方法。

1.1 资源获取
Getting pytest

pytest 的官方文档地址：https://docs.pytest.org。pytest 通过 PyPI（Python 官方包管理索引）分发托管：https://pypi.python.org/pypi/pytest。

就像其他在 PyPI 中托管的 Python 程序包一样，你可以在当前的虚拟环境中使用 pip 安装 pytest：

```
$ pip3 install -U virtualenv
$ python3 -m virtualenv venv
$ source venv/bin/activate
$ pip install pytest
```

如果你不太熟悉 vritualenv 或者 pip，请查阅附录 A。

> **Windows、Python 2、venv 的问题**
>
> virtualenv 和 pip 的例子应该可以在 Linux、macOS 等 POSIX（可移植操作系统标准接口）系统下正常运行，也支持多个 Python 版本，包括 2.7.9 以后的所有版本。
>
> 但是在 Windows 下，source/venv/bin/activate 这条命令无法使用，请使用 venv\Scripts\activate.bat：
>
> ```
> C:\> pip3 install -U virtualenv
> C:\> python3 -m virtualenv venv
> C:\>venv\Scripts\activate.bat (venv)
> C:\> pip install pytest
> ```
>
> Python 3.6 以后的版本最好使用 venv 替代 virtualenv。Python 3.6 自带 venv，不必手动安装。但我听说在某些平台下，virtualenv 的表现更好。

1.2 运行 Pytest
Running pytest

```
$ pytest --help
usage: pytest [options] [file_or_dir] [file_or_dir] [...]
    ...
```

如果你不提供任何参数，pytest 会在当前目录以及子目录下寻找测试文件，然后运行搜索到的测试代码。如果你提供一个或多个文件名、目录名，pytest 会逐个查找并运行所有测试。为了搜索到所有的测试代码，pytest 会递归遍历每个目录及其子目录。

举一个例子，我们新建一个 tasks 子目录，并且创建以下测试文件：

```
ch1/tasks/test_three.py
"""Test the Task data type."""
from collections import namedtuple
Task = namedtuple('Task', ['summary', 'owner', 'done', 'id'])
Task.__new__.__defaults__ = (None, None, False, None)

def test_defaults():
    """Using no parameters should invoke defaults."""
    t1 = Task()
    t2 = Task(None, None, False, None)
    assert t1 == t2

def test_member_access():
    """Check .field functionality of namedtuple."""
    t = Task('buy milk', 'brian')
    assert t.summary == 'buy milk'
    assert t.owner == 'brian'
    assert (t.done, t.id) == (False, None)
```

你可以使用 __new__.__defaults__ 创建默认的 Task 对象，不必指定所有属性。测试用例 test_defaults() 中演示了默认值的校验。

测试用例 test_member_access() 演示了如何利用属性名（而不是索引）来访问对象成员，这也是选用 namedtuple 的一个原因。

下面两个例子演示了 _asdict() 函数和 _replace() 函数的功能：

```
cha1/tasks/test_four.py
"""Test the Task data type."""

from collections import namedtuple

Task = namedtuple('Task', ['summary', 'owner', 'done', 'id'])
Task.__new__.__defaults__ = (None, None, False, None)

def test_asdict():
    """_asdict() should return a dictionary."""
    t_task = Task('do something', 'okken', True, 21)
    t_dict = t_task._asdict()
    expected = {'summary': 'do something',
                'owner': 'okken',
                'done': True,
                'id': 21}
    assert t_dict == expected

def test_replace():
    """replace() should change passed in fields."""
    t_before = Task('finish book', 'brian', False)
    t_after = t_before._replace(id=10, done=True)
    t_expected = Task('finish book', 'brian', True, 10)
    assert t_after == t_expected
```

运行 pytest 时可以指定目录和文件。如果不指定，pytest 会搜索当前目录及其子目录中以 test_开头或以_test 结尾的测试函数。假设切换到 ch1 目录运行 pytest（没有指定任何参数），那么 pytest 会运行 ch1 目录下的四个测试文件。

```
$ cd /path/to/code/ch1
$ pytest
===================== test session starts =====================
collected 6 items

test_one.py .
test_two.py F
tasks/test_four.py ..
tasks/test_three.py ..
=========================== FAILURES ==========================
_____ test_failing _____
    def test_failing():
>       assert (1, 2, 3) == (3, 2, 1)
E       assert (1, 2, 3) == (3, 2, 1)
E         At index 0 diff: 1 != 3
E         Use -v to get the full diff
test_two.py:2: AssertionError
============== 1 failed, 5 passed in 0.08 seconds =============
```

如果我们编写了新的测试，那么可以在 pytest 中指定需要测试的文件名或目录，或者预先切换到需要运行的目录：

```
$ pytest tasks/test_three.py tasks/test_four.py
=================== test session starts ====================
collected 4 items

tasks/test_three.py ..
tasks/test_four.py ..
================= 4 passed in 0.02 seconds =================
$ pytest tasks
=================== test session starts ====================
collected 4 items

tasks/test_four.py ..
tasks/test_three.py ..
================= 4 passed in 0.03 seconds =================
$ cd /path/to/code/ch1/tasks
$ pytest
=================== test session starts ====================
collected 4 items

test_four.py ..
test_three.py ..
================= 4 passed in 0.02 seconds =================
```

我们把 pytest 搜索测试文件和测试用例的过程称为测试搜索（test discovery）。只要你遵守 pytest 的命名规则，pytest 就能自动搜索所有待执行的测试用例。以下是几条主要的命名规则。

- 测试文件应当命名为 `test_<something>.py` 或者 `<something>_test.py`。
- 测试函数、测试类方法应当命名为 `test_<something>`。
- 测试类应当命名为 `Test<Something>`。

测试文件和测试函数最好以 `test_` 开头，但如果先前编写的测试用例遵循的是其他命名规则，也可以修改默认的测试搜索规则，第 6 章会介绍这方面的技巧。

下面我们来逐句讲解 pytest 运行单个测试文件时的控制台输出信息：

```
$ cd /path/to/code/ch1/tasks
$ pytest test_three.py
================= test session starts ==================
platform darwin -- Python 3.6.2, pytest-3.2.1, py-1.4.34, pluggy-0.4.0
rootdir: /path/to/code/ch1/tasks, inifile:
collected 2 items

test_three.py ..
=============== 2 passed in 0.01 seconds ===============
```

从中我们可以获得很多信息：

===== *test session starts* ====

> pytest 为每段测试会话（session）做了明确的分隔，一段会话就是 pytest 的一次调用，它可能包括多个目录下被执行的测试用例。后面还会介绍到，如果 pytest fixture 的作用范围是会话级别，那么会话的定义就显得尤为重要（参见第 3.6 节）。

platform darwin -- Python 3.6.2, pytest-3.2.1, py-1.4.34, pluggy-0.4.0

> 我使用的是 Mac，所以显示的是 `platform darwin`，Windows 的表示方式略有不同。接着显示的是 Python、pytest 以及 pytest 包的版本。py 和 pluggy 都是 pytest 包，用于 pytest 的实现，它们均由 pytest 团队开发维护。

rootdir: /path/to/code/ch1/tasks, inifile:

> `rootdir`（当前起始目录）是 pytest 搜索测试代码时最常使用的目录，`inifile` 用于列举配置文件（这里没有指定），文件名可能是 `pytest.ini`、`tox.ini` 或者 `setup.cfg`。关于配置文件的知识，请参考第 6 章。

collected 2 items

> 搜索范围内找到两个测试条目。

`test_three.py ..`

three.py 表示测试文件，每个文件的测试情况只占据一行，两个点号表示两个测试用例均已通过。点号仅仅表示通过，而 Failure（失败）、error（异常）、skip（跳过）、xfail（预期失败）、xpass（预期失败但通过）会被分别标记为 F、E、s、x、X。使用-v 或--verbose 可以看到更多细节。

`== 2 passed in 0.01 seconds ==`

表示测试通过的数量以及这段会话耗费的时间，如果存在未通过的测试用例，则会根据未通过的类型列举数量。

测试结果是测试人员了解测试始末的主要途径。在 pytest 中，测试函数可能返回多种结果，不只是通过或失败。

以下是可能出现的类型。

- PASSED(.)：测试通过。
- FAILED(F)：测试失败（也有可能是 XPASS 状态与 strict 选项冲突造成的失败，见后文）。
- SKIPPED(s)：测试未被执行。指定测试跳过执行，可以将测试标记为 `@pytest.mark.skip()`，或者使用`@pytest.mark.skipif()`指定跳过测试的条件，请参考第 2.5 节。
- xfail(x)：预期测试失败，并且确实失败。使用`@pytest.mark.xfail()`指定你认为会失败的测试用例，请参考第 2.6 节。
- XPASS(X)：预期测试失败，但实际上运行通过，不符合预期。
- ERROR(E)：测试用例之外的代码触发了异常，可能由 fixture 引起，也可能由 hook 函数引起。

1.3 运行单个测试用例
Running Only One Test

学习测试最好从单个测试用例开始。你可以直接在指定文件后方添加::test_name，像下面这样：

```
$ cd /path/to/code/ch1
$ pytest -v tasks/test_four.py::test_asdict
=================== test session starts ===================
collected 3 items

tasks/test_four.py::test_asdict PASSED
================ 1 passed in 0.01 seconds =================
```

接下来看命令行选项。

1.4 使用命令行选项
Using Options

我们已经使用过-v 和--verbose 选项。pytest 还提供很多选项，本书只会用到一部分。你可以使用 pytest--help 查看全部选项。

下面列出常用的 pytest 命令选项，对 pytest 初学者而言，这些选项已经够用了。

```
$ pytest --help
  ... subset of the list ...
  -k EXPRESSION         only run tests/classes which match the given
                        substring expression.
                        Example: -k 'test_method or test_other' matches
                        all test functions and classes whose name
                        contains 'test_method' or 'test_other'.
  -m MARKEXPR           only run tests matching given mark expression.
                        example: -m 'mark1 and not mark2'.
  -x, --exitfirst       exit instantly on first error or failed test.
  --maxfail=num         exit after first num failures or errors.
  --capture=method      per-test capturing method: one of fd|sys|no.
  -s                    shortcut for --capture=no.
  --lf, --last-failed   rerun only the tests that failed last time
                        (or all if none failed)
```

```
--ff, --failed-first  run all tests but run the last failures first.
-v, --verbose         increase verbosity.
-q, --quiet           decrease verbosity.
-l, --showlocals      show locals in tracebacks (disabled by default).
--tb=style            traceback print mode (auto/long/short/line/native/no).
--durations=N         show N slowest setup/test durations (N=0 for all).
--collect-only        only collect tests, don't execute them.
--version             display pytest lib version and import information.
-h, --help            show help message and configuration info
```

--collect-only 选项

--collect-only

使用--collect-only 选项可以展示在给定的配置下哪些测试用例会被运行。之所以首先介绍它，是因为该选项的输出内容可以作为一种参照。打开 ch1 目录，使用此选项，你将看到本章到目前为止使用的全部测试用例。

```
$ cd /path/to/code/ch1
$ pytest --collect-only
=================== test session starts ===================
collected 6 items
 <Module 'test_one.py'>
  <Function 'test_passing'>
 <Module 'test_two.py'>
  <Function 'test_failing'>
 <Module 'tasks/test_four.py'>
  <Function 'test_asdict'>
  <Function 'test_replace'>
 <Module 'tasks/test_three.py'>
  <Function 'test_defaults'>
  <Function 'test_member_access'>
============== no tests ran in 0.03 seconds ===============
```

--collect-only 选项可以让你非常方便地在测试运行之前，检查选中的测试用例是否符合预期。下面介绍-k 选项时，我们还会用到它。

-k 选项

-k EXPRESSION

-k 选项允许你使用表达式指定希望运行的测试用例。这个功能非常实用，如果某测试名是唯一的，或者多个测试名的前缀或后缀相同，那么可以使用表达式来快速定位。假设希望选中 test_asdict() 和 test_defaults()，那么可以使用 --collect-only 验证筛选情况：

```
$ cd /path/to/code/ch1
$ pytest -k "asdict or defaults" --collect-only
=================== test session starts ===================
collected 6 items
<Module 'tasks/test_four.py'>
  <Function 'test_asdict'>
<Module 'tasks/test_three.py'>
  <Function 'test_defaults'>
=================== 4 tests deselected ===================
============== 4 deselected in 0.03 seconds ==============
```

看起来符合预期。现在把 --collect-only 选项移除，让它们正常运行。

```
$ pytest -k "asdict or defaults"
=================== test session starts ===================
collected 6 items

tasks/test_four.py .
tasks/test_three.py .
=================== 4 tests deselected ===================
========= 2 passed, 4 deselected in 0.03 seconds ==========
```

很好，它们都通过了。不过运行结果也符合我们的预期吗？可以打开 -v 或者 --verbose 查看：

```
$ pytest -v -k "asdict or defaults"
=================== test session starts ===================
collected 6 items

tasks/test_four.py::test_asdict PASSED
tasks/test_three.py::test_defaults PASSED
=================== 4 tests deselected ===================
========= 2 passed, 4 deselected in 0.02 seconds ==========
```

很好，两个用例都通过验证了。

-m 选项
-m MARKEXPR

标记（marker）用于标记测试并分组，以便快速选中并运行。以

test_replace()和 test_member_access()为例，它们甚至都不在同一个文件里，如果你希望同时选中它们，那么可以预先做好标记。

使用什么标记名由你自己决定，假设你希望使用 run_these_please，则可以使用@pytest.mark.run_these_please 这样的装饰器（decorator）来做标记，像下面这样：

```
import pytest
...
@pytest.mark.run_these_please
def test_member_access():
...
```

给 test_replace()也做上同样的标记。有相同标记的测试（集合），可以一起运行。例如，这里我们使用 pytest -m run_these_please 命令就可以同时运行 test_replace()和 test_member_access()。

```
$ cd /path/to/code/ch1/tasks
$ pytest -v -m run_these_please
================= test session starts ==================
collected 4 items

test_four.py::test_replace PASSED
test_three.py::test_member_access PASSED
================== 2 tests deselected ==================
========= 2 passed, 2 deselected in 0.02 seconds =========
```

使用-m 选项可以用表达式指定多个标记名。使用-m "mark1 and mark2"可以同时选中带有这两个标记的所有测试用例。使用-m "mark1 and not mark2" 则会选中带有 mark1 的测试用例，而过滤掉带有 mark2 的测试用例；使用-m "mark1 or mark2"则选中带有 mark1 或者 mark2 的所有测试用例。第 2.4 节还会详细介绍测试函数的标记。

-x 选项
-x, --exitfirst

正常情况下，pytest 会运行每一个搜索到的测试用例。如果某个测试函数被断言失败，或者触发了外部异常，则该测试用例的运行就会到此为止，pytest 将其标记为失败后会继续运行下一个测试用例。通常，这就是我们期望的运行模式。但是在 debug 时，我们会希望遇到失败时立即停止整个会话，这时 -x 选项就派上用场了。

让我们试着用 -x 选项运行之前的 6 个测试用例：

```
$ cd /path/to/code/ch1
$ pytest -x
=================== test session starts ===================
collected 6 items

test_one.py .
test_two.py F
========================= FAILURES =========================
_____ test_failing _____
    def test_failing():
>       assert (1, 2, 3) == (3, 2, 1)
E       assert (1, 2, 3) == (3, 2, 1)
E         At index 0 diff: 1 != 3
E         Use -v to get the full diff
test_two.py:2: AssertionError
!!!!!!!!! Interrupted: stopping after 1 failures !!!!!!!!!!
=========== 1 failed, 1 passed in 0.25 seconds ============
```

输出信息开头显示 pytest 收集到了 6 个测试条目，末尾显示有 1 个通过，有 1 个失败。Interrupted 提示我们测试中断了。

如果没有 -x 选项，那么 6 个测试都会被执行，去掉 -x 再运行一次，并且使用 --tb=no 选项关闭错误信息回溯：

```
$ cd /path/to/code/ch1
$ pytest --tb=no
================== test session starts ==================
collected 6 items
test_one.py .
test_two.py F
tasks/test_four.py ..
tasks/test_three.py ..
=========== 1 failed, 5 passed in 0.09 seconds ============
```

可以看到，不使用-x 时，pytest 在 test_two.py 文件中遇到了测试失败，但并没有停止执行后面的测试用例。

--maxfail=num

--maxfail=num

-x 选项的特点是，一旦遇到测试失败，就会全局停止。假设你允许 pytest 失败几次后再停止，则可以使用--maxfail 选项，明确指定可以失败几次。

以目前仅仅存在一个失败测试的情况，很难展示这个特性，如果我们设置--maxfail=2，那么所有的测试都会被运行；如果设置--maxfail=1，就与-x 的作用相同。下面让我们来试试。

```
$ cd /path/to/code/ch1
$ pytest --maxfail=2 --tb=no
================== test session starts ==================
collected 6 items

test_one.py .
test_two.py F
tasks/test_four.py ..
tasks/test_three.py ..
=========== 1 failed, 5 passed in 0.08 seconds ============
$ pytest --maxfail=1 --tb=no
================== test session starts ==================
collected 6 items

test_one.py .
test_two.py F
!!!!!!!!!! Interrupted: stopping after 1 failures !!!!!!!!!!
=========== 1 failed, 1 passed in 0.19 seconds ============
```

这一次我们也使用--tb=no 关闭了错误堆栈回溯。

-s 与 --capture=method
-s and --capture=method

-s 选项允许终端在测试运行时输出某些结果，包括任何符合标准的输出流信息。-s 等价于--capture=no。正常情况下，所有的测试输出都会被捕获。测试失败时，为了帮助你理解究竟发生了什么，pytest 会做出推断，并输出报告。-st 和--capture=no 选项关闭了输出捕获。我编写测试用例时，习惯添加几个 print()，以便观察某时刻测试执行到了哪个阶段。

使用-l/--showlocals 选项，在测试失败时会打印出局部变量名和它们的值，这样可以规避一些不必要的 print 语句。

信息捕获方法还有--capture=fd 和--capture=sys。使用--capture=sys 时，sys.stdout/stderr 将被输出至内存；使用--capture=fd 时，若文件描述符（file descriptor）为 1 或 2，则会被输出至临时文件中。

实际上，我自己很少使用 sys 和 fd，我用得最多的是-s。因为讲到-s，所以不得不提到--capture。

目前为止，我们的例子中还没有使用过打印语句，但是我建议读者自己尝试一下。

--lf (--last-failed) 选项
-lf, --last-failed

当一个或多个测试失败时，我们常常希望能够定位到最后一个失败的测试用例重新运行，这时可以使用--lf 选项。

```
$ cd /path/to/code/ch1
$ pytest --lf
=================== test session starts ===================
run-last-failure: rerun last 1 failures
collected 6 items

test_two.py F
========================= FAILURES ========================
```

```
_____ test_failing _____
    def test_failing():
>       assert (1, 2, 3) == (3, 2, 1)
E       assert (1, 2, 3) == (3, 2, 1)
E         At index 0 diff: 1 != 3
E         Use -v to get the full diff
test_two.py:2: AssertionError
=================== 5 tests deselected ====================
========== 1 failed, 5 deselected in 0.08 seconds =========
```

这里最好加上 --tb 选项，可以隐藏部分信息，或者指定其他的错误堆栈回溯模式，重新运行失败的测试用例。

--ff（--failed-first）选项
-ff, --failed-first

--ff（--failed-first）选项与 --last-failed 选项的作用基本相同，不同之处在于 --ff 会运行完剩余的测试用例。

```
$ cd /path/to/code/ch1
$ pytest --ff --tb=no
==================== test session starts ==================
run-last-failure: rerun last 1 failures first
collected 6 items

test_two.py F
test_one.py .
tasks/test_four.py ..
tasks/test_three.py ..
=========== 1 failed, 5 passed in 0.09 seconds ============
```

由于 test_failing() 是在 test_two.py 文件中，因此通常会在 test_one.py 之后运行，但在 --ff 选项作用下，test_failing() 前一轮被认定为失败，会被首先执行。

-v（--verbose）选项
-v, --verbose

使用 -v/--verbose 选项，输出的信息会更详细。最明显的区别就是每个文件中的每个测试用例都占一行（先前是每个文件占一行），测试的名字和结果

都会显示出来,而不仅仅是一个点或字符。我们已经多次使用过-v,现在可以在--ff中的例子添加该选项,与上文作对比:

```
$ cd /path/to/code/ch1
$ pytest -v --ff --tb=no
=================== test session starts ===================
run-last-failure: rerun last 1 failures first
collected 6 items

test_two.py::test_failing FAILED
test_one.py::test_passing PASSED
tasks/test_four.py::test_asdict PASSED
tasks/test_four.py::test_replace PASSED
tasks/test_three.py::test_defaults PASSED
tasks/test_three.py::test_member_access PASSED
=========== 1 failed, 5 passed in 0.07 seconds ============
```

在彩色显示器上可以看到FAILED标记为红色,PASSED标记为绿色。

-q (--quiet) 选项

-q, --quiet

该选项的作用与-v/--verbose 的相反,它会简化输出信息。我喜欢将-q和--tb=line(仅打印异常的代码位置)搭配使用。

单独使用-q的情况:

```
$ cd /path/to/code/ch1
$ pytest -q .F....
========================= FAILURES =========================
_____ test_failing _____
    def test_failing():
>       assert (1, 2, 3) == (3, 2, 1)
E       assert (1, 2, 3) == (3, 2, 1)
E         At index 0 diff: 1 != 3
E         Full diff:
E         - (1, 2, 3)
E         ?    ^    ^
E         + (3, 2, 1)
E         ?    ^    ^

test_two.py:2: AssertionError
1 failed, 5 passed in 0.08 seconds
```

使用-q 选项会简化输出信息，只保留最核心的内容。为了节省篇幅和突出重点，后文会频繁使用-q 和--tb=no。

-l (--showlocals) 选项

-l, --showlocals

使用-l 选项，失败测试用例由于被堆栈追踪，所以局部变量及其值都会显示出来。

目前为止，我们还没有包含局部变量的失败测试，因此我对 test_replace() 做了一些修改。

t_expected = Task(*'finish book'*, *'brian'*, True, 10)

改为：

t_expected = Task(*'finish book'*, *'brian'*, True, 11)

10 改成 11，由于 11 不是测试函数期望的值，所以会造成测试失败。这样就可以使用-l/--showlocals 选项了。

```
$ cd /path/to/code/ch1
$ pytest -l tasks
=================== test session starts ===================
collected 4 items

tasks/test_four.py .F
tasks/test_three.py ..
========================= FAILURES ========================
_____ test_replace _____
    def test_replace():
        t_before = Task('finish book', 'brian', False)
        t_after = t_before._replace(id=10, done=True)
        t_expected = Task('finish book', 'brian', True, 11)
>       assert t_after == t_expected
E       AssertionError: assert Task(summary=...e=True, id=10) == Task(summary='...e=True, id=11)
E         At index 3 diff: 10 != 11
E         Use -v to get the full diff
t_after= Task(summary='finish book', owner='brian', done=True, id=10)
t_before= Task(summary='finish book', owner='brian', done=False, id=None)
t_expected = Task(summary='finish book', owner='brian', done=True,
```

```
    id=11)
tasks/test_four.py:20: AssertionError
=========== 1 failed, 3 passed in 0.08 seconds ============
```

assert 触发测试失败之后，代码片断下方显示的是本地变量 t_after、t_beforeh、t_expected 详细的值。

--tb=style 选项

--tb=style

--tb=style 选项决定捕捉到失败时输出信息的显示方式。某个测试用例失败后，pytest 会列举出失败信息，包括失败出现在哪一行、是什么失败、怎么失败，此过程我们称之为"信息回溯"。大多数情况下，信息回溯是必要的，它对找到问题很有帮助，但有时也会对多余的信息感到厌烦，这时--tb=style 选项就有用武之地了。我推荐的 style 类型有 short、line、no。short 模式仅输出 assert 的一行以及系统判定内容（不显示上下文）；line 模式只使用一行输出显示所有的错误信息；no 模式则直接屏蔽全部回溯信息。

继续使用上文中对于 test_replace() 的修改，利用它来看看各种失败回溯信息的显示方式。

使用--tb=no 屏蔽全部回溯信息。

```
$ cd /path/to/code/ch1
$ pytest --tb=no tasks
=================== test session starts ===================
collected 4 items

tasks/test_four.py .F
tasks/test_three.py ..
=========== 1 failed, 3 passed in 0.04 seconds ============
```

使用--tb=line，它可以告诉我们错误的位置，有时这已经足够了（特别是在运行大量测试用例的情况下，可以用它发现失败的共性）。

```
$ pytest --tb=line tasks
=================== test session starts ===================
collected 4 items
```

```
tasks/test_four.py .F
tasks/test_three.py ..
======================= FAILURES ========================
/path/to/code/ch1/tasks/test_four.py:20:
AssertionError: assert Task(summary=...e=True, id=10) == Task(
summary='...e=True, id=11)
=========== 1 failed, 3 passed in 0.05 seconds ============
```

使用--tb=short，显示的回溯信息比前面两种模式的更详细。

```
$ pytest --tb=short tasks
================== test session starts ==================
collected 4 items

tasks/test_four.py .F
tasks/test_three.py ..
======================= FAILURES ========================
_____ test_replace _____
tasks/test_four.py:20: in test_replace
    assert t_after == t_expected
E   AssertionError: assert Task(summary=...e=True, id=10) == Task(
summary='...e=True, id=11)
E     At index 3 diff: 10 != 11
E     Use -v to get the full diff
=========== 1 failed, 3 passed in 0.04 seconds ============
```

这些信息足够让你了解发生了什么。

除此之外，还有三种可选的模式。

--tb=long 输出最为详尽的回溯信息；--tb=auto 是默认值，如果有多个测试用例失败，仅打印第一个和最后一个用例的回溯信息（格式与 long 模式的一致）；--tb=native 只输出 Python 标准库的回溯信息，不显示额外信息。

--duration=N 选项
--durations=N

--duration=N选项可以加快测试节奏。它不关心测试是如何运行的，只统计测试过程中哪几个阶段是最慢的（包括每个测试用例的 call、setup、teardown 过程）。它会显示最慢的 N 个阶段，耗时越长越靠前。如果使用

--duration=0，则会将所有阶段按耗时从长到短排序后显示。

本书的例子耗时都比较短，为了方便演示，我特意添加了一个休眠函数 time.sleep(0.1)，猜猜它在哪里。

```
$ cd /path/to/code/ch1
$ pytest --durations=3 tasks
================= test session starts =================
collected 4 items

tasks/test_four.py ..
tasks/test_three.py ..
============== slowest 3 test durations ===============
0.10s call     tasks/test_four.py::test_replace
0.00s setup    tasks/test_three.py::test_defaults
0.00s teardown tasks/test_three.py::test_member_access
============== 4 passed in 0.13 seconds ===============
```

最慢的阶段出现了，它的标签是 call，显然是它被休眠了 0.1 秒。紧随其后的两个阶段是 setup 和 teardown。每个测试用例大体上都包含三个阶段：call、setup、teardown。其中 setup 和 teardown 也称 fixture，你可以在 fixture 中添加代码，在测试之前让系统做好准备，在测试之后做必要的清理。第 3 章会详细讲解 fixture。

--version 选项

--version

使用--version 可以显示当前的 pytest 版本及安装目录：

```
$ pytest --version
This is pytest version 3.0.7, imported from
 /path/to/venv/lib/python3.5/site-packages/pytest.py
```

由于是在虚拟环境中安装的 pytest，所以会定位到对应的 site-packages 目录。

-h (--help) 选项
-h, --help

即使你已经能够熟练使用 pytest，-h 选项依然非常有用，它不但能展示原生 pytest 的用法，还能展示新添加的插件的选项和用法。

使用 -h 选项可以获得：

- 基本用法：pytest [options] [file_or_dir] [file_or_dir] [...]

- 命令行选项及其用法，包括新添加的插件的选项及其用法。

- 可用于 ini 配置文件中的选项（第 6 章会详细介绍）。

- 影响 pytest 行为的环境变量（第 6 章会详细介绍）。

- 使用 pytest --markers 时的可用 marker 列表（第 2 章会详细介绍）。

- 使用 pytest --fixtures 时的可用 fixture 列表（第 3 章会详细介绍）。

帮助信息最后会显示一句话：

shown according to specified file_or_dir or current dir if not specified

这句话的意思为：显示结果取决于指定的文件或目录，未指定的则默认使用当前目录和文件。这句话非常重要，选项、marker、fixture 都会因为目录的变化而发生变化。这是因为 pytest 有可能在某个叫 conftest.py 的文件中搜索到 hook 函数（新增命令行选项）、新的 fixture 定义及新的 marker 定义。

Pytest 允许你在 conftest.py 和测试文件中自定义测试行为，其作用域仅仅是某个项目，甚至只是某个项目的测试子集。第 6 章会详细介绍 conftest.py 和 ini 配置文件。

1.5 练习
Exercises

1. 使用 python -m virtualenv 或 python -m venv 创建一个新的虚拟环境，专供本书的示例使用。我以前也拒绝这么做，但目前已经离不开它了。如果遇到困难，可以参见附录 A。

2. 尝试进入和退出虚拟环境。

   - `$ source venv/bin/activate`
   - `$ deactivate`

   Windows 环境：

   - `C:\Users\okken\sandbox>venv\scripts\activate.bat`
   - `C:\Users\okken\sandbox>deactivate`

3. 在虚拟环境中安装 pytest，如果遇困难，请参考附录 B。即使你已经安装过 pytest，也请你在虚拟环境中再安装一次。

4. 创建几个测试文件，可以借鉴前文的示例。使用 pytest 运行这几个测试文件。

5. 更改 assert 语句，不要使用 assert something == something_else 这样的语句，尝试使用类似下面这样的例子：

   - `assert 1 in [2, 3, 4]`
   - `assert a < b`
   - `assert 'fizz' not in 'fizzbuzz'`

1.6 预告
What's Next

本章介绍了 pytest 的获取及执行方式,但还没有介绍如何编写测试函数。第 2 章将讲解如何编写测试函数,如何使用不同的参数调用测试函数,如何将测试分发到类、模块和组件包。

第 2 章

编写测试函数
Writing Test Functions

在第 1 章中,我们学习了 pytest 是如何工作的,学习了如何指定测试目录、使用命令行选项。本章将讲解如何为 Python 程序包编写测试函数。即使你要测试的不是 Python 包,本章的大部分内容仍然适用。

在为 Tasks 项目编写测试之前,我会先介绍典型的可分发 Python 包的目录结构。然后演示如何在测试中使用 assert,如何处理可预期和不可预期的异常。

我会讲解如何借助类、模块、目录来组织测试,以便管理大量的测试。还会学习使用 marker 来标记希望同时运行的测试,使用内置 marker 跳过某些测试,为预期会失败的测试做标记。最后将介绍参数化测试,以便使用多组数据开展测试。

2.1 测试示例程序
Testing a Package

下面将以 Tasks 项目来演示如何为 Python 程序包编写测试。Tasks 是一个包含同名命令行工具的 Python 程序包。

附录 D 介绍了如何使用 PyPI 在团队内部或互联网上分发你的项目，在此不多赘述。下面来看看 Tasks 项目的目录结构，请思考各个文件在测试过程中所起到的作用。

以下是 Tasks 项目的文件结构：

```
tasks_proj/
├── CHANGELOG.rst
├── LICENSE
├── MANIFEST.in
├── README.rst
├── setup.py
├── src
│   └── tasks
│       ├── __init__.py
│       ├── api.py
│       ├── cli.py
│       ├── config.py
│       ├── tasksdb_pymongo.py
│       └── tasksdb_tinydb.py
└── tests
    ├── conftest.py
    ├── pytest.ini
    ├── func
    │   ├── __init__.py
    │   ├── test_add.py
    │   └── ...
    └── unit
        ├── __init__.py
        ├── test_task.py
        └── ...
```

这个目录结构展示了测试文件与整个项目的关系，其中有几个关键的文件值得我们注意：conftest.py、pytest.ini、__init__.py、setup.py，它们将在测试中发挥重要作用。

所有的测试都放在 tests 文件夹里，与包含程序源码的 src 文件夹分开。这并非 pytest 的硬性要求，而是一个很好的习惯。

根目录文件 CHANGELOG.rst、LICENSE、README.rst、MANIFEST.in、

setup.py 将在附录 D 中详细介绍。setup.py 对于创建可分发的包很重要，本地其他项目可以将它作为依赖包导入。

功能测试和单元测试放在不同的目录下，这也不是硬性规定，但这样做可以让你更方便地分别运行两类测试。我习惯将这两类测试分开放，因为功能测试只会在改变系统功能时才有可能发生中断异常，而单元测试的中断在代码重构、业务逻辑实现期间都有可能发生。

项目目录中包含两类 __init__.py 文件，一类出现在 src 目录下，一类出现在 tests 目录下。src/tasks/__init__.py 告诉 Python 解释器该目录是 Python 包。此外，执行到 import tasks 时，这个 __init__.py 将作为该包主入口。该文件包含导入 api.py 模块的代码，因此 api.py 中的函数可以直接被 cli.py 和测试模块访问，比如可以直接使用 task.add()，而不必使用 task.api.add()。

test/func/__init__.py 和 test/unit/__init__.py 都是空文件，它们的作用是给 pytest 提供搜索路径，找到测试根目录以及 pytest.ini 文件。

pytest.ini 文件是可选的，它保存了 pytest 在该项目下的特定配置。项目中顶多包含一个配置文件，其中的指令可以调节 pytest 的工作行为，例如配置常用的命令行选项列表。第 6 章会详细介绍 pytest.ini。

conftest.py 文件同样是可选的，它是 pytest 的"本地插件库"，其中可以包含 hook 函数和 fixture。hook 函数可以将自定义逻辑引入 pytest，用于改善 pytest 的执行流程；fixture 则是一些用于测试前后执行配置及销毁逻辑的外壳函数，可以传递测试中用到的资源。第 3 章和第 4 章会详细介绍 fixtures。第 5 章会详细讲解 hook 函数。在包含多个子目录的测试中，hook 函数和 fixtures 的使用将被定义在 tests/conftests.py 内。同一个项目内可以包含多个 conftest.py 文件，例如 tests 目录下可以有一个 conftest.py 文件，在 tests 的每个子目录下也可以各有一个 conftest.py 文件。

如果你的测试目录没有按照规范来建立，可以在本书网站下载该项目的源码[1]，在后续工作中，可以按照类似结构存放测试文件。

本地安装 Tasks 项目程序包
Installing a Package Locally

测试文件 tests/test_task.py 中包含第 1.2 节用到的测试用例（原来存放在 test_three.py 和 test_four.py 文件里）。现在将这几个函数都归并到同一个文件里，并且使用了更有意义的名字。另外，还删除了原先文件中对 Task 数据结构的定义，它应该写在 api.py 里。

以下是 test_tasks.py:

```
ch2/tasks_proj/tests/unit/test_tasks.py
"""Test the Task data type."""
from tasks import Task

def test_asdict():
    """_asdict() should return a dictionary."""
    t_task = Task('do something', 'okken', True, 21)
    t_dict = t_task._asdict()
    expected = {'summary': 'do something',
                'owner': 'okken',
                'done': True, 'id': 21}
    assert t_dict == expected

def test_replace():
    """replace() should change passed in fields."""
    t_before = Task('finish book', 'brian', False)
    t_after = t_before._replace(id=10, done=True)
    t_expected = Task('finish book', 'brian', True, 10)
    assert t_after == t_expected

def test_defaults():
    """Using no parameters should invoke defaults."""
    t1 = Task()
    t2 = Task(None, None, False, None)
    assert t1 == t2
```

---

[1] https://pragprog.com/titles/bopytest/source_code

```python
def test_member_access():
    """Check .field functionality of namedtuple."""
    t = Task('buy milk', 'brian')
    assert t.summary == 'buy milk'
    assert t.owner == 'brian'
    assert (t.done, t.id) == (False, None)
```

test_task.py 文件中包含以下 import 语句:

```python
from tasks import Task
```

测试中如果希望使用 import tasks 或 from tasks import something,最好的方式是在本地使用 pip 安装 tasks 包。包的根目录包含一个 setup.py 文件, pip 可以直接使用。

你可以切换到 tasks_proj 根目录,运行 pip install . 或 pip install -e . 来安装 tasks 包,也可以在上层目录使用 pip install -e task_proj 来安装 tasks 包。

```
$ cd /path/to/code
$ pip install ./tasks_proj/
$ pip install --no-cache-dir ./tasks_proj/
Processing ./tasks_proj
Collecting click (from tasks==0.1.0)
  Downloading click-6.7-py2.py3-none-any.whl (71kB)
    ...
Collecting tinydb (from tasks==0.1.0)
  Downloading tinydb-3.4.0.tar.gz
Collecting six (from tasks==0.1.0)
  Downloading six-1.10.0-py2.py3-none-any.whl
Installing collected packages: click, tinydb, six, tasks
  Running setup.py install for tinydb ... done
  Running setup.py install for tasks ... done
Successfully installed click-6.7 six-1.10.0 tasks-0.1.0 tinydb-3.4.0
```

如果仅仅是测试 tasks 包,这个命令就足够了,但是,如果安装后希望修改源码重新安装,就需要使用 -e 选项(editable)。

```
$ pip install -e ./tasks_proj/
Obtaining file:///path/to/code/tasks_proj Requirement already
satisfied: click in
  /path/to/venv/lib/python3.6/site-packages (from tasks==0.1.0)
Requirement already satisfied: tinydb in
```

```
  /path/to/venv/lib/python3.6/site-packages (from tasks==0.1.0)
Requirement already satisfied: six in
  /path/to/venv/lib/python3.6/site-packages (from tasks==0.1.0)
Installing collected packages: tasks
  Found existing installation: tasks 0.1.0
    Uninstalling tasks-0.1.0:
      Successfully uninstalled tasks-0.1.0
  Running setup.py develop for tasks
Successfully installed tasks
```

现在来运行测试:

```
$ cd /path/to/code/ch2/tasks_proj/tests/unit
$ pytest test_task.py
===================== test session starts =====================
collected 4 items

test_task.py ....
=================== 4 passed in 0.01 seconds ==================
```

import 语句执行成功，说明 tasks 包已正常安装。接下来的测试就可以使用 import tasks 了，下面来编写一些测试。

2.2 使用 assert 声明
Using assert Statements

用 pytest 编写测试时，若需要传递测试失败信息，则可以直接使用 Python 自带的 assert 关键字，这样做很方便，也是许多开发者选择 pytest 的原因。

如果使用其他测试框架，则可能会看到许多以 assert 开头的函数。下面列举了使用 assert 与各种以 assert 开头的函数的区别:

```
pytest                  unittest
assert something        assertTrue(something)
assert a == b           assertEqual(a, b)
assert a <= b           assertLessEqual(a, b)
...                     ...
```

pytest 允许在 assert 关键字后面添加任何表达式(assert <expression>)。如果表达式的值通过 bool 转换后等于 False，则意味着测试失败。

2.2 使用 assert 声明

pytest 有一个重要功能是可以重写 assert 关键字。pytest 会截断对原生 assert 的调用，替换为 pytest 定义的 assert，从而提供更多的失败信息和细节。在下面的例子中，可以看到重写 assert 关键字是多么必要：

ch2/tasks_proj/tests/unit/test_task_fail.py
```python
"""Use the Task type to show test failures."""
from tasks import Task

def test_task_equality():
    """Different tasks should not be equal."""
    t1 = Task('sit there', 'brian')
    t2 = Task('do something', 'okken')
    assert t1 == t2

def test_dict_equality():
    """Different tasks compared as dicts should not be equal."""
    t1_dict = Task('make sandwich', 'okken')._asdict()
    t2_dict = Task('make sandwich', 'okkem')._asdict()
    assert t1_dict == t2_dict
```

以上都是 assert 失败的例子，可以看到回溯的信息很丰富。

```
$ cd /path/to/code/ch2/tasks_proj/tests/unit
$ pytest test_task_fail.py
===================== test session starts =====================
collected 2 items

test_task_fail.py FF
=========================== FAILURES ==========================
_____ test_task_equality _____
    def test_task_equality():
        t1 = Task('sit there', 'brian')
        t2 = Task('do something', 'okken')
>       assert t1 == t2
E       AssertionError: assert Task(summary=...alse, id=None) == Task(summary='...alse, id=None)
E         At index 0 diff: 'sit there' != 'do something'
E         Use -v to get the full diff
test_task_fail.py:6: AssertionError
_____ test_dict_equality _____
    def test_dict_equality():
        t1_dict = Task('make sandwich', 'okken')._asdict()
        t2_dict = Task('make sandwich', 'okkem')._asdict()
>       assert t1_dict == t2_dict
E       AssertionError: assert OrderedDict([...('id', None)]) == OrderedDict([[(...('id', None)])
```

```
E         Omitting 3 identical items, use -v to show
E         Differing items:
E         {'owner': 'okken'} != {'owner': 'okkem'}
E         Use -v to get the full diff
test_task_fail.py:11: AssertionError
=================== 2 failed in 0.06 seconds ===================
```

上面每个失败的测试用例在行首都用一个>号来标识。以 E 开头的行是 pytest 提供的额外判定信息，用于帮助我们了解异常的具体情况。

我有意将两列不匹配的参数放到测试用例 test_task_equality()中，但我们看到 pytest 只展示了第一列结果。按照错误信息中给出的建议，我们使用-v 选项再执行一遍：

```
$ pytest -v test_task_fail.py::test_task_equality
===================== test session starts =====================
collected 3 items

test_task_fail.py::test_task_equality FAILED
========================== FAILURES ===========================
_____ test_task_equality _____
    def test_task_equality():
        t1 = Task('sit there', 'brian')
        t2 = Task('do something', 'okken')
>       assert t1 == t2
E       AssertionError: assert Task(summary=...alse, id=None) == Task(summary='...alse, id=None)
E         At index 0 diff: 'sit there' != 'do something'
E         Full diff:
E         - Task(summary='sit there', owner='brian', done=False, id=None)
E         ?            ^^^  ^^^  ^^^^
E         + Task(summary='do something', owner='okken', done=False, id=None)
E         ?            +++^^^  ^^^  ^^^^
test_task_fail.py:6: AssertionError
=================== 1 failed in 0.07 seconds ===================
```

这样就更一目了然了。pytest 不单指出了区别，还给出了区别详情。

这个例子中的 assert 仅用于判定是否相等。你可以在 pytest.org 上找到很多更复杂的 assert 例子。

2.3 预期异常
Expecting Exceptions

在 Tasks 项目的 API 中，有几个地方可能抛出异常。tasks/api.py 中有以下几个函数。

```
def add(task):  # type: (Task) -> int
def get(task_id):  # type: (int) -> Task
def list_tasks(owner=None):  # type: (str|None) -> list of Task
def count():  # type: (None) -> int
def update(task_id, task):  # type: (int, Task) -> None
def delete(task_id):  # type: (int) -> None
def delete_all():  # type: () -> None
def unique_id():  # type: () -> int
def start_tasks_db(db_path, db_type):  # type: (str, str) -> None
def stop_tasks_db():  # type: () -> None
```

cli.py 中的 CLI 代码与 api.py 中的 API 代码统一指定了发送给 API 函数的数据类型，假设检查到数据类型错误，异常很可能是由这些 API 函数抛出的。

为确保这些函数在发生类型错误时可以抛出异常，下面来做一下检验：在测试中使用错误类型的数据，引起 TypeError 异常。同时，我还使用了 with pytest.raises(<expected exception>)声明，像这样：

```
ch2/tasks_proj/tests/func/test_api_exceptions.py
import pytest
import tasks


def test_add_raises():
    """add() should raise an exception with wrong type param."""
    with pytest.raises(TypeError):
        tasks.add(task='not a Task object')
```

测试用例 test_add_raises()中有 with pytest.raises(TypeError)声明，意味着无论 with 中的内容是什么，都至少会发生 TypeError 异常。如果测试通过，说明确实发生了我们预期的 TypeError 异常；如果抛出的是其他类型的异常，则与我们所预期的不一致，说明测试失败。

上面的测试中只检验了传参数据的"类型异常",你也可以检验"值异常"。对于 start_tasks_db(db_path, db_type) 来说,db_type 不单要求是字符串类型,还必须为 'tiny' 或 'mongo'。为校验异常消息是否符合预期,可以通过增加 as excinfo 语句得到异常消息的值,再进行比对。

ch2/tasks_proj/tests/func/test_api_exceptions.py
```python
def test_start_tasks_db_raises():
    """Make sure unsupported db raises an exception."""
    with pytest.raises(ValueError) as excinfo:
        tasks.start_tasks_db('some/great/path', 'mysql')
    exception_msg = excinfo.value.args[0]
    assert exception_msg == "db_type must be a 'tiny' or 'mongo'"
```

现在可以更细致地观察异常。as 后面的变量是 ExceptionInfo 类型,它会被赋予异常消息的值。

在这个例子中,我们希望确保第一个(也是唯一的)参数引起的异常消息能与某个字符串匹配。

2.4 测试函数的标记
Marking Test Functions

pytest 提供了标记机制,允许你使用 marker 对测试函数做标记。一个测试函数可以有多个 marker,一个 marker 也可以用来标记多个测试函数。

讲解 marker 的作用需要看实际的例子。比如我们选一部分测试作为冒烟测试,以了解当前系统中是否存在大的缺陷。通常冒烟测试不会包含全套测试,只选择可以快速出结果的测试子集,让开发者对系统健康状况有一个大致的了解。

为了把选定的测试加入冒烟测试,可以对它们添加 @pytest.mark.smoke 装饰器。先用 test_api_exceptions.py 中的两个测试作为例子(注意:smoke 和 get 标记是我定义的,并非 pytest 内置的)。

ch2/tasks_proj/tests/func/test_api_exceptions.py
```python
@pytest.mark.smoke
def test_list_raises():
    """List() should raise an exception with wrong type param."""
    with pytest.raises(TypeError):
        tasks.list_tasks(owner=123)

@pytest.mark.get
@pytest.mark.smoke
def test_get_raises():
    """get() should raise an exception with wrong type param."""
    with pytest.raises(TypeError):
        tasks.get(task_id='123')
```

现在只需要在命令中指定-m marker_name，就可以运行它们。

```
$ cd /path/to/code/ch2/tasks_proj/tests/func
$ pytest -v -m 'smoke' test_api_exceptions.py
===================== test session starts =====================
collected 7 items

test_api_exceptions.py::test_list_raises PASSED
test_api_exceptions.py::test_get_raises PASSED
===================== 5 tests deselected ======================
============ 2 passed, 5 deselected in 0.03 seconds ===========
$ pytest -v -m 'get' test_api_exceptions.py
===================== test session starts =====================
collected 7 items

test_api_exceptions.py::test_get_raises PASSED
===================== 6 tests deselected ======================
============ 1 passed, 6 deselected in 0.01 seconds ===========
```

别忘了-v 是--verbose 的简写，它用于展示具体运行的测试的名字。指定-m 'smoke'，会运行标记为@pytest.mark.smoke 的两个测试；而指定-m 'get'，会运行标记为@pytest.mark.get 的那个测试。这样做显得清晰明了。

-m 后面也可以使用表达式，可以在标记之间添加 and、or、not 关键字。

```
$ pytest -v -m 'smoke and get' test_api_exceptions.py
===================== test session starts =====================
collected 7 items
test_api_exceptions.py::test_get_raises PASSED
===================== 6 tests deselected ======================
============ 1 passed, 6 deselected in 0.03 seconds ===========
```

上面的命令只会运行既有 smoke 标记，又有 get 标记的测试。还可以加上 not。

```
$ pytest -v -m 'smoke and not get' test_api_exceptions.py
===================== test session starts =====================
collected 7 items
test_api_exceptions.py::test_list_raises PASSED
===================== 6 tests deselected =====================
============ 1 passed, 6 deselected in 0.03 seconds ============
```

'smoke and not get' 的作用是筛选出有 smoke 标记，但没有 get 标记的测试。

完善冒烟测试
Filling Out the Smoke Test

上面的例子还不能算是合理的冒烟测试，因为两个测试并未涉及数据库改动（显然这是有必要的）。

现在来编写两个测试，它们都包含增加 task 对象的逻辑，然后把其中一个放到冒烟测试里。

```
ch2/tasks_proj/tests/func/test_add.py
import pytest
import tasks
from tasks import Task

def test_add_returns_valid_id():
    """tasks.add(<valid task>) should return an integer."""
    # GIVEN an initialized tasks db
    # WHEN a new task is added
    # THEN returned task_id is of type int
    new_task = Task('do something')
    task_id = tasks.add(new_task)
    assert isinstance(task_id, int)

@pytest.mark.smoke
def test_added_task_has_id_set():
    """Make sure the task_id field is set by tasks.add()."""
    # GIVEN an initialized tasks db
    # AND a new task is added
```

```
new_task = Task('sit in chair', owner='me', done=True)
task_id = tasks.add(new_task)

# WHEN task is retrieved
task_from_db = tasks.get(task_id)

# THEN task_id matches id field
assert task_from_db.id == task_id
```

两个测试中都包含一句注释：GIVEN an initialized tasks db，但实际上并未涉及数据库初始化。现在这里可以定义一个 fixture，用于测试前后控制数据库的连接。

```
ch2/tasks_proj/tests/func/test_add.py
@pytest.fixture(autouse=True)
def initialized_tasks_db(tmpdir):
    """Connect to db before testing, disconnect after."""
    # Setup : start db
    tasks.start_tasks_db(str(tmpdir), 'tiny')

    yield # this is where the testing happens

    # Teardown : stop db
    tasks.stop_tasks_db()
```

这个例子中使用的 tmpdir 是 pytest 内置的 fixture。第 4 章将详细介绍内置 fixture，第 3 章将讲解如何编写及使用自定义的 fixture，以及这里用到的 autouse 参数。

本例中的 autouse 表示当前文件中的所有测试都将使用该 fixture。yield 之前的代码将在测试运行前执行，而 yield 之后的代码会在测试运行后执行。若有必要，yield 也可以返回数据给测试。后续章节会详细介绍 fixture，这里由于需要初始化数据库，才提前使用它。（pytest 也支持经典的 nose 和 unittest 风格的 setup 和 teardown 函数。感兴趣的读者可以查阅附录 F。）

我们暂时把对 fixture 的讨论放到一边。请切换到项目根目录，运行我们的冒烟测试。

```
$ cd /path/to/code/ch2/tasks_proj
$ pytest -v -m 'smoke'
===================== test session starts ======================
collected 56 items

tests/func/test_add.py::test_added_task_has_id_set PASSED
tests/func/test_api_exceptions.py::test_list_raises PASSED
tests/func/test_api_exceptions.py::test_get_raises PASSED
===================== 53 tests deselected ======================
=========== 3 passed, 53 deselected in 0.11 seconds ============
```

可以看到，带有相同标记的测试即使存放在不同的文件下，也会被一起执行。

2.5 跳过测试
Skipping Tests

第 2.4 节使用的标记是自定义的。pytest 自身内置了一些标记：`skip`、`skipif`、`xfail`。本节介绍 `skip` 和 `skipif`，下一节介绍 `xfail`。

`skip` 和 `skipif` 允许你跳过不希望运行的测试。比方，我们不确定 `tasks.unique_id()` 是否会按照预期的方式工作：它每次调用后返回的是不同的数值吗？它会返回一个数据库中尚不存在的值吗？

为此，下面来编写一个测试（注意上一节的 `initialized_tasks_db` fixture 仍然有效）。

```
ch2/tasks_proj/tests/func/test_unique_id_1.py
import pytest
import tasks
def test_unique_id():
    """Calling unique_id() twice should return different numbers."""
    id_1 = tasks.unique_id()
    id_2 = tasks.unique_id()
    assert id_1 != id_2
```

然后执行这个测试。

```
$ cd /path/to/code/ch2/tasks_proj/tests/func
$ pytest test_unique_id_1.py
===================== test session starts =====================
collected 1 item s

test_unique_id_1.py F
========================== FAILURES ===========================
_____ test_unique_id _____
    def test_unique_id():
        """Calling unique_id() twice should return different numbers."""
        id_1 = tasks.unique_id()
        id_2 = tasks.unique_id()
>       assert id_1 != id_2
E       assert 1 != 1
test_unique_id_1.py:12: AssertionError
==================== 1 failed in 0.06 seconds ====================
```

很遗憾，我们的测试方法有问题。查看 API 文档后，发现其中有一句话：`"""Return an integer that does not exist in the db."""`（会返回一个数据库中不存在的整型数，但是没有说会保证每次返回值都不同。）

我们可以直接修改测试，不过让我们试试把它标记为skip。

ch2/tasks_proj/tests/func/test_unique_id_2.py
```
@pytest.mark.skip(reason='misunderstood the API')
def test_unique_id_1():
    """Calling unique_id() twice should return different numbers."""
    id_1 = tasks.unique_id()
    id_2 = tasks.unique_id()
    assert id_1 != id_2

def test_unique_id_2():
    """unique_id() should return an unused id."""
    ids = []
    ids.append(tasks.add(Task('one')))
    ids.append(tasks.add(Task('two')))
    ids.append(tasks.add(Task('three')))
    # grab a unique id
    uid = tasks.unique_id()
    # make sure it isn't in the list of existing ids
    assert uid not in ids
```

要跳过某个测试，只需要简单地在测试函数上方添加@pytest.mark.skip()装饰器即可。

再运行一次。

```
$ pytest -v test_unique_id_2.py
===================== test session starts =========================
collected 2 items

test_unique_id_2.py::test_unique_id_1 SKIPPED
test_unique_id_2.py::test_unique_id_2 PASSED
==============1 passed, 1 skipped in 0.02 seconds ====================
```

实际上，可以给要跳过的测试添加理由和条件，比如希望它只在包版本低于 0.2.0 时才生效，这时应当使用 skipif 来替代 skip：

ch2/tasks_proj/tests/func/test_unique_id_3.py
```
@pytest.mark.skipif(tasks.__version__ < '0.2.0',
                    reason='not supported until version 0.2.0')

def test_unique_id_1():
    """Calling unique_id() twice should return different numbers."""
    id_1 = tasks.unique_id()
    id_2 = tasks.unique_id()
    assert id_1 != id_2
```

skipif() 的判定条件可以是任何 Python 表达式，这里比对的是包版本。

无论是 skip 还是 skipif，我都写上了跳过的理由（尽管 skip 并不要求这样做）。我习惯在使用 skip、skipif、xfail 时都写清楚理由。

以下是修改后的测试输出情况。

```
$ pytest test_unique_id_3.py
===================== test session starts =========================
collected 2 items

test_unique_id_3.py s.
============ 1 passed, 1 skipped in 0.02 seconds ====================
```

s. 表明有一个测试被跳过，有一个测试通过。使用 -v 选项可以看到具体是哪一个测试被跳过，哪一个测试通过了。

```
$ pytest -v test_unique_id_3.py
===================== test session starts =========================
collected 2 items
```

```
test_unique_id_3.py::test_unique_id_1 SKIPPED
test_unique_id_3.py::test_unique_id_2 PASSED
============= 1 passed, 1 skipped in 0.03 seconds ==================
```

但我们仍然看不到跳过的原因，这时可以使用 -rs。

```
$ pytest -rs test_unique_id_3.py
======================== test session starts =========================
collected 2 items

test_unique_id_3.py s.
====================== short test summary info ======================
SKIP [1] func/test_unique_id_3.py:5: not supported until version 0.2.0
================ 1 passed, 1 skipped in 0.03 seconds ================
```

-r chars 选项有帮助文档。

```
$ pytest --help
...
  -r chars

    show extra test summary info as specified by chars
    (f)ailed, (E)error, (s)skipped, (x)failed, (X)passed,
    (p)passed, (P)passed with output, (a)all except pP.
...
```

该选项不但可以帮助用户了解某些测试被跳过的原因，还可以用于查看其他测试结果。

2.6 标记预期会失败的测试
Marking Tests as Expecting to Fail

使用 skip 和 skipif 标记，测试会直接跳过，而不会被执行。使用 xfail 标记，则告诉 pytest 运行此测试，但我们预期它会失败。接下来，我会使用 xfail 修改 unique_id() 测试：

```
ch2/tasks_proj/tests/func/test_unique_id_4.py
@pytest.mark.xfail(tasks.__version__ < '0.2.0',
                   reason='not supported until version 0.2.0')
def test_unique_id_1():
```

```python
    """Calling unique_id() twice should return different numbers."""
    id_1 = tasks.unique_id()
    id_2 = tasks.unique_id()
    assert id_1 != id_2

@pytest.mark.xfail()
def test_unique_id_is_a_duck():
    """Demonstrate xfail."""
    uid = tasks.unique_id()
    assert uid == 'a duck'

@pytest.mark.xfail()
def test_unique_id_not_a_duck():
    """Demonstrate xpass."""
    uid = tasks.unique_id()
    assert uid != 'a duck'
```

第一个测试与上一节的一样，只不过我用 xfail 替换了 skip。后面两个测试都使用了 xfail 标记，区别在于一个是==，一个是!=，所以结果必然是一个通过，一个失败。

下面执行该测试文件。

```
$ cd /path/to/code/ch2/tasks_proj/tests/func
$ pytest test_unique_id_4.py
======================= test session starts ========================
collected 4 items

test_unique_id_4.py xxX.
=========== 1 passed, 2 xfailed, 1 xpassed in 0.07 seconds ===========
```

x 代表 XFAIL，意味着"expected to fail"（预期失败，实际上也失败了）。大写的 X 代表 XPASS，意味着"expected to fail but passed"（预期失败，但实际运行并没有失败）。

使用--verbose 选项，将逐条输出更多信息。

```
$ pytest -v test_unique_id_4.py
======================= test session starts ========================
collected 4 items

test_unique_id_4.py::test_unique_id_1 xfail
test_unique_id_4.py::test_unique_id_is_a_duck xfail
test_unique_id_4.py::test_unique_id_not_a_duck XPASS
```

```
test_unique_id_4.py::test_unique_id_2 PASSED
========== 1 passed, 2 xfailed, 1 xpassed in 0.08 seconds ===========
```

对于标记为 xfail，但实际运行结果是 XPASS 的测试，可以在 pytest 配置中强制指定结果为 FAIL（像下面这样修改 pytest.ini 文件即可）。

```
[pytest]
xfail_strict=true
```

第 6 章将详细介绍 pytest.ini 文件。

2.7 运行测试子集
Running a Subset of Tests

前面已经讨论了如何给测试做标记，以及如何根据标记运行测试。运行测试子集有很多种方式，不但可以选择运行某个目录、文件、类中的测试，还可以选择运行某一个测试用例（可能在文件中，也可能在类中）。本节将介绍测试类，并通过表达式来完成测试函数名的字符串匹配。

单个目录
A Single Directory

运行单个目录下的所有测试，以目录作为 pytest 的参数即可。

```
$ cd /path/to/code/ch2/tasks_proj
$ pytest tests/func --tb=no
===================== test session starts =====================
collected 50 items

tests/func/test_add.py ..
tests/func/test_add_variety.py ................................
tests/func/test_api_exceptions.py .......
tests/func/test_unique_id_1.py F
tests/func/test_unique_id_2.py s.
tests/func/test_unique_id_3.py s.
tests/func/test_unique_id_4.py xxX.
 1 failed, 44 passed, 2 skipped, 2 xfailed, 1 xpassed in 0.26 seconds
```

加上 -v 选项后再运行一次。通过观察其中的命名输出方式，学习如何指定目录、类、测试。

```
$ pytest -v tests/func --tb=no
===================== test session starts ======================
collected 50 items

tests/func/test_add.py::test_add_returns_valid_id PASSED
tests/func/test_add.py::test_added_task_has_id_set PASSED
...
tests/func/test_api_exceptions.py::test_add_raises PASSED
tests/func/test_api_exceptions.py::test_list_raises PASSED
tests/func/test_api_exceptions.py::test_get_raises PASSED
...
tests/func/test_unique_id_1.py::test_unique_id FAILED
tests/func/test_unique_id_2.py::test_unique_id_1 SKIPPED
tests/func/test_unique_id_2.py::test_unique_id_2 PASSED
...
tests/func/test_unique_id_4.py::test_unique_id_1 xfail
tests/func/test_unique_id_4.py::test_unique_id_is_a_duck xfail
tests/func/test_unique_id_4.py::test_unique_id_not_a_duck XPASS
tests/func/test_unique_id_4.py::test_unique_id_2 PASSED

1 failed, 44 passed, 2 skipped, 2 xfailed, 1 xpassed in 0.30 seconds
```

下面几个例子会用到上面的命名方式。

单个测试文件/模块
A Single Test File/Module

运行单个文件里的全部测试，以路径名加文件名作为 pytest 参数即可。

```
$ cd /path/to/code/ch2/tasks_proj
$ pytest tests/func/test_add.py
===================== test session starts ==========================
collected 2 items

tests/func/test_add.py ..
================== 2 passed in 0.05 seconds =========================
```

这样的例子前面已经出现许多次了。

单个测试函数
A Single Test Function

运行单个测试函数，只需要在文件名后面添加::符号和函数名。

```
$ cd /path/to/code/ch2/tasks_proj
$ pytest -v tests/func/test_add.py::test_add_returns_valid_id
=================== test session starts ==========================
collected 3 items

tests/func/test_add.py::test_add_returns_valid_id PASSED
================ 1 passed in 0.02 seconds ========================
```

使用-v可以显示执行的是哪个函数。

单个测试类
A Single Test Class

测试类用于将某些相似的测试函数组合在一起，举个例子：

ch2/tasks_proj/tests/func/test_api_exceptions.py
```
class TestUpdate():
    """Test expected exceptions with tasks.update()."""

    def test_bad_id(self):
        """A non-int id should raise an excption."""
        with pytest.raises(TypeError):
            tasks.update(task_id={'dict instead': 1},
                         task=tasks.Task())

    def test_bad_task(self):
        """A non-Task task should raise an excption."""
        with pytest.raises(TypeError):
            tasks.update(task_id=1, task='not a task')
```

由于这两个测试都是在测试 update()函数，把它们放到一个类中是合理的。要运行该类，可以在文件名后面加上::符号和类名（与运行单个测试函数类似）。

```
$ cd /path/to/code/ch2/tasks_proj
$ pytest -v tests/func/test_api_exceptions.py::TestUpdate
=================== test session starts ==========================
collected 7 items
```

```
tests/func/test_api_exceptions.py::TestUpdate::test_bad_id PASSED
tests/func/test_api_exceptions.py::TestUpdate::test_bad_task PASSED
=================== 2 passed in 0.03 seconds ==========================
```

单个测试类中的测试方法
A Single Test Method of a Test Class

如果不希望运行测试类中的所有测试，只想指定运行其中一个，一样可以在文件名后面添加::符号和方法名。

```
$ cd /path/to/code/ch2/tasks_proj
$ pytest -v tests/func/test_api_exceptions.py::TestUpdate::test_bad_id
====================== test session starts ======================
collected 1 item

tests/func/test_api_exceptions.py::TestUpdate::test_bad_id PASSED
=================== 1 passed in 0.03 seconds ===================
```

> **通过列举详情了解句法**
>
> 注意：运行测试子集时，不必记忆指定目录、文件、函数、类、类方法的句法，其格式与 pytest -v 的输出是一致的。

用测试名划分测试集合
A Set of Tests Based on Test Name

-k 选项允许用一个表达式指定需要运行的测试，该表达式可以匹配测试名（或其子串）。表达式中可以包含 and、or、not。

下面来运行所有名字中包含 _raises 的测试。

```
$ cd /path/to/code/ch2/tasks_proj
$ pytest -v -k _raises
====================== test session starts ======================
collected 56 items
```

```
tests/func/test_api_exceptions.py::test_add_raises PASSED
tests/func/test_api_exceptions.py::test_list_raises PASSED
tests/func/test_api_exceptions.py::test_get_raises PASSED
tests/func/test_api_exceptions.py::test_delete_raises PASSED
tests/func/test_api_exceptions.py::test_start_tasks_db_raises PASSED
===================== 51 tests deselected ======================
=========== 5 passed, 51 deselected in 0.07 seconds ============
```

如果要跳过 test_delete_raises() 的执行，则可以使用 and 和 not。

```
$ pytest -v -k "_raises and not delete"
===================== test session starts ======================
collected 56 items

tests/func/test_api_exceptions.py::test_add_raises PASSED
tests/func/test_api_exceptions.py::test_list_raises PASSED
tests/func/test_api_exceptions.py::test_get_raises PASSED
tests/func/test_api_exceptions.py::test_start_tasks_db_raises PASSED
===================== 52 tests deselected ======================
=========== 4 passed, 52 deselected in 0.06 seconds ============
```

上面学习了如何指定测试文件、目录、类、函数，如何利用 -k 选项选择部分测试。下面介绍如何将一个测试函数用作多个测试用例，即以多组测试数据运行。

2.8 参数化测试
Parametrized Testing

向函数传值并检验输出结果是软件测试的常用手段，但是对大部分功能测试而言，仅仅使用一组数据是无法充分测试函数功能的。参数化测试允许传递多组数据，一旦发现测试失败，pytest 会及时报告。

为了介绍参数化测试需要解决的问题，我举一个关于 add() 的例子。

```
ch2/tasks_proj/tests/func/test_add_variety.py
import pytest
import tasks
from tasks import Task
```

```python
def test_add_1():
    """tasks.get() using id returned from add() works."""
    task = Task('breathe', 'BRIAN', True)
    task_id = tasks.add(task)
    t_from_db = tasks.get(task_id)
    # everything but the id should be the same
    assert equivalent(t_from_db, task)

    def equivalent(t1, t2):
        """Check two tasks for equivalence."""
        # Compare everything but the id field
        return ((t1.summary == t2.summary) and
                (t1.owner == t2.owner) and
                (t1.done == t2.done))

@pytest.fixture(autouse=True)
def initialized_tasks_db(tmpdir):
    """Connect to db before testing, disconnect after."""
    tasks.start_tasks_db(str(tmpdir), 'tiny')
    yield
    tasks.stop_tasks_db()
```

新创建的 Task 对象，其 id 会被置为 None，只有在发生数据库交互之后其 id 才会被填入相应的值，因此，要比较两个 Task 对象的值是否相等，不能使用==，而应该使用 equivalent()方法。equivalent()方法检查除 id 之外的所有属性是否相等。autouse 的 fixture 能确保数据库可访问。下面来看看测试是否能通过。

```
$ cd /path/to/code/ch2/tasks_proj/tests/func
$ pytest -v test_add_variety.py::test_add_1
===================== test session starts =====================
collected 1 item

test_add_variety.py::test_add_1 PASSED
================== 1 passed in 0.03 seconds ===================
```

测试通过了，但只测了一个 task。能不能批量测试呢？当然可以。可以使用@pytest.mark.parametrize(argnames, argvalues)装饰器达到批量传送参数的目的。

ch2/tasks_proj/tests/func/test_add_variety.py
```python
@pytest.mark.parametrize('task',
                         [Task('sleep', done=True),
                          Task('wake', 'brian'),
                          Task('breathe', 'BRIAN', True),
                          Task('exercise', 'BrIaN', False)])
def test_add_2(task):
    """Demonstrate parametrize with one parameter."""
    task_id = tasks.add(task)
    t_from_db = tasks.get(task_id)
    assert equivalent(t_from_db, task)
```

parametrize()的第一个参数是用逗号分隔的字符串列表，本例中只有一个'task'；第二个参数是一个值列表，本例中是一个 Task 对象列表。pytest 会轮流对每个 task 做测试，并分别报告每一个测试用例的结果。

```
$ cd /path/to/code/ch2/tasks_proj/tests/func
$ pytest -v test_add_variety.py::test_add_2
===================== test session starts =====================
collected 4 items

test_add_variety.py::test_add_2[task0] PASSED
test_add_variety.py::test_add_2[task1] PASSED
test_add_variety.py::test_add_2[task2] PASSED
test_add_variety.py::test_add_2[task3] PASSED
=================== 4 passed in 0.05 seconds ==================
```

parametrize()函数工作正常，再来看看把 task 对象列表替换为几组键值对会如何。

ch2/tasks_proj/tests/func/test_add_variety.py
```python
@pytest.mark.parametrize('summary, owner, done',
                         [('sleep', None, False),
                          ('wake', 'brian', False),
                          ('breathe', 'BRIAN', True),
                          ('eat eggs', 'BrIaN', False),
                          ])
def test_add_3(summary, owner, done):
    """Demonstrate parametrize with multiple parameters."""
    task = Task(summary, owner, done)
    task_id = tasks.add(task)
    t_from_db = tasks.get(task_id)
    assert equivalent(t_from_db, task)
```

如果你使用了易于被 pytest 转换为字符串的数据对象，则输出结果中的测试标识会使用组合后的参数值以增强可读性。

```
$ cd /path/to/code/ch2/tasks_proj/tests/func
$ pytest -v test_add_variety.py::test_add_3
===================== test session starts =====================
collected 4 items

test_add_variety.py::test_add_3[sleep-None-False] PASSED
test_add_variety.py::test_add_3[wake-brian-False] PASSED
test_add_variety.py::test_add_3[breathe-BRIAN-True] PASSED
test_add_variety.py::test_add_3[eat eggs-BrIaN-False] PASSED
================== 4 passed in 0.05 seconds ===================
```

你可以使用完整的测试标识（pytest 术语为 node）来重新指定需要运行的测试。

```
$ cd /path/to/code/ch2/tasks_proj/tests/func
$ pytest -v test_add_variety.py::test_add_3[sleep-None-False]
===================== test session starts =====================
collected 1 item

test_add_variety.py::test_add_3[sleep-None-False] PASSED
================== 1 passed in 0.02 seconds ===================
```

如果标识中包含空格，别忘了添加引号。

```
$ cd /path/to/code/ch2/tasks_proj/tests/func
$ pytest -v "test_add_variety.py::test_add_3[eat eggs-BrIaN-False]"
===================== test session starts =====================
collected 1 item

test_add_variety.py::test_add_3[eat eggs-BrIaN-False] PASSED
================== 1 passed in 0.03 seconds ===================
```

现在回头看看参数为 task 对象列表的情况，下面先将 task 列表移出装饰器。

```
ch2/tasks_proj/tests/func/test_add_variety.py
tasks_to_try = (Task('sleep', done=True),
                Task('wake', 'brian'),
                Task('wake', 'brian'),
                Task('breathe', 'BRIAN', True),
                Task('exercise', 'BrIaN', False))
```

```python
@pytest.mark.parametrize('task', tasks_to_try)
def test_add_4(task):
    """Slightly different take."""
    task_id = tasks.add(task)
    t_from_db = tasks.get(task_id)
    assert equivalent(t_from_db, task)
```

这样代码看起来比较美观，但测试输出信息的可读性较差。

```
$ cd /path/to/code/ch2/tasks_proj/tests/func
$ pytest -v test_add_variety.py::test_add_4
===================== test session starts =====================
collected 5 items

test_add_variety.py::test_add_4[task0] PASSED
test_add_variety.py::test_add_4[task1] PASSED
test_add_variety.py::test_add_4[task2] PASSED
test_add_variety.py::test_add_4[task3] PASSED
test_add_variety.py::test_add_4[task4] PASSED
=================== 5 passed in 0.05 seconds ==================
```

在多参数情况下，测试输出的可读性非常好，但这里只显示了 Task 对象列表。为了改善可读性，我们为 parametrize()引入一个额外参数 ids，使列表中的每一个元素都被标识。ids 是一个字符串列表，它和数据对象列表的长度保持一致。由于给数据集分配了一个变量 tasks_to_try，所以可以通过它生成 ids。

ch2/tasks_proj/tests/func/test_add_variety.py
```python
task_ids = ['Task({},{},{})'.format(t.summary, t.owner, t.done)
            for t in tasks_to_try]

@pytest.mark.parametrize('task', tasks_to_try, ids=task_ids)
def test_add_5(task):
    """Demonstrate ids."""
    task_id = tasks.add(task)
    t_from_db = tasks.get(task_id)
    assert equivalent(t_from_db, task)
```

下面来运行。

```
$ cd /path/to/code/ch2/tasks_proj/tests/func
$ pytest -v test_add_variety.py::test_add_5
===================== test session starts =====================
```

```
collected 5 items

test_add_variety.py::test_add_5[Task(sleep,None,True)] PASSED
test_add_variety.py::test_add_5[Task(wake,brian,False)0] PASSED
test_add_variety.py::test_add_5[Task(wake,brian,False)1] PASSED
test_add_variety.py::test_add_5[Task(breathe,BRIAN,True)] PASSED
test_add_variety.py::test_add_5[Task(exercise,BrIaN,False)] PASSED
=================== 5 passed in 0.04 seconds ===================
```

现在，自定义测试标识已经能够被 pytest 识别了。

```
$ cd /path/to/code/ch2/tasks_proj/tests/func
$ pytest -v
"test_add_variety.py::test_add_5[Task(exercise,BrIaN,False)]"
===================== test session starts =====================
collected 1 item

test_add_variety.py::test_add_5[Task(exercise,BrIaN,False)] PASSED
=================== 1 passed in 0.03 seconds ===================
```

建议为这些标识加上引号，否则方括号和圆括号在命令行中会显得难以理解。

你也可以为测试类加上 parametrize()装饰器，这种情况下，该数据集会被传递给该类的所有类方法。

ch2/tasks_proj/tests/func/test_add_variety.py
```python
@pytest.mark.parametrize('task', tasks_to_try, ids=task_ids)
class TestAdd():
    """Demonstrate parametrize and test classes."""

    def test_equivalent(self, task):
        """Similar test, just within a class."""
        task_id = tasks.add(task)
        t_from_db = tasks.get(task_id)
        assert equivalent(t_from_db, task)

    def test_valid_id(self, task):
        """We can use the same data or multiple tests."""
        task_id = tasks.add(task)
        t_from_db = tasks.get(task_id)
        assert t_from_db.id == task_id
```

以下是执行结果。

```
$ cd /path/to/code/ch2/tasks_proj/tests/func
$ pytest -v test_add_variety.py::TestAdd
===================== test session starts ======================
collected 10 items

test_add_variety.py::TestAdd::test_equivalent[Task(sleep,None,True)]
PASSED
test_add_variety.py::TestAdd::test_equivalent[Task(wake,brian,False)0]
PASSED
test_add_variety.py::TestAdd::test_equivalent[Task(wake,brian,False)1]
PASSED
test_add_variety.py::TestAdd::test_equivalent[Task(breathe,BRIAN,True)]
PASSED
test_add_variety.py::TestAdd::test_equivalent[Task(exercise,BrIaN,False
)] PASSED
test_add_variety.py::TestAdd::test_valid_id[Task(sleep,None,True)]
PASSED
test_add_variety.py::TestAdd::test_valid_id[Task(wake,brian,False)0]
PASSED
test_add_variety.py::TestAdd::test_valid_id[Task(wake,brian,False)1]
PASSED
test_add_variety.py::TestAdd::test_valid_id[Task(breathe,BRIAN,True)]
PASSED
test_add_variety.py::TestAdd::test_valid_id[Task(exercise,BrIaN,False)]
PASSED
================== 10 passed in 0.08 seconds ===================
```

在给@pytest.mark.parametrize()装饰器传入列表参数时，还可以在参数值旁边定义一个 id 来做标识，语法是 pytest.param(<value>, id="something")。

ch2/tasks_proj/tests/func/test_add_variety.py
```
@pytest.mark.parametrize('task', [
    pytest.param(Task('create'), id='just summary'),
    pytest.param(Task('inspire', 'Michelle'), id='summary/owner'),
    pytest.param(Task('encourage', 'Michelle', True),
id='summary/owner/done')])
def test_add_6(task):
    """Demonstrate pytest.param and id."""
    task_id = tasks.add(task)
```

```
t_from_db = tasks.get(task_id)
assert equivalent(t_from_db, task)
```

以下是运行情况。

```
$ cd /path/to/code/ch2/tasks_proj/tests/func
$ pytest -v test_add_variety.py::test_add_6
=================== test session starts ====================
collected 3 items

test_add_variety.py::test_add_6[just summary] PASSED
test_add_variety.py::test_add_6[summary/owner] PASSED
test_add_variety.py::test_add_6[summary/owner/done] PASSED
================= 3 passed in 0.05 seconds =================
```

在 id 不能被参数化批量生成，需要自定义的时候，这个方法很管用。

2.9 练习
Exercises

1. 访问本书主页，下载本章中的示例项目 `tasks_proj`，使用 `pip install /path/to/tasks_proj` 完成本地安装。

2. 仔细查看 tests 目录。

3. 以单个文件运行 pytest。

4. 在单目录下运行 pytest，例如 `tasks_proj/tests/func`。使用 pytest 单独运行该目录下某个指定的测试，或同时运行该目录下的全部测试。其中会有失败的测试用例，想想它们为什么失败。

5. 使用 pytest 直接运行 `tests` 目录时，给失败的测试用例添加 `xfail` 或 `skip` 标记，直到结果中不存在测试失败。

6. 目前还没有测试所有 API 函数，包括 `tasks.count()`。挑选一个待测的 API 函数，思考为了保证它工作正常，必须编写哪些测试用例。

7. 如果你添加的 task 对象的 id 已经存在，那该怎么办？你可以在 test_api_exceptions.py 文件中添加一个用于验证该异常的测试用例。这个小练习可以参考 tasks_proj 中的 api.py。

2.10 预告
What's Next

我相信读到这里，pytest 已经能够极大地简化你的测试工作了。本章多次使用了名为 initialized_tasks_db 的 fixture，fixture 的作用是将一些非核心测试逻辑（如测试数据的检索和生成）从测试函数中分离出来，以便于其他测试函数复用，同时保持这些边缘逻辑的一致性。第 3 章将详细介绍 fixture。

第 3 章

pytest Fixture

我们已经学习了 pytest 的基础知识,接下来将介绍 fixture。测试任何规模稍大的软件系统,fixture 都可以派上用场。fixture 是在测试函数运行前后,由 pytest 执行的外壳函数。fixture 中的代码可以定制,满足多变的测试需求,包括定义传入测试中的数据集、配置测试前系统的初始状态、为批量测试提供数据源,等等。

以下是一个返回数值的简单 fixture。

```
ch3/test_fixtures.py
import pytest

@pytest.fixture()
def some_data():
    """Return answer to ultimate question."""
    return 42

def test_some_data(some_data):
    """Use fixture return value in a test."""
    assert some_data == 42
```

@pytest.fixture()装饰器用于声明函数是一个 fixture。如果测试函数的参数列表中包含 fixture 名,那么 pytest 会检测到,并在测试函数运行之前执行该 fixture。fixture 可以完成任务,也可以返回数据给测试函数。

测试用例 test_some_data() 的参数列表中包含一个 fixture 名 some_data，pytest 会以该名称搜索 fixture（可见命名在 pytest 中是非常重要的）。pytest 会优先搜索该测试所在的模块，然后搜索 conftest.py。

在开始介绍 fixture 和 conftest.py 之前，必须指出：fixture 作为编程领域、测试领域，甚至是 Python 语言层面的专业术语，是存在多种含义的。本章中使用的 fixture、fixture 函数、fixture 方法都是指由@pytest.fixture()装饰器定义的函数。有时 fixture 也可以指由 fixture 函数配置的程序资源。fixture 函数常用于配置、检索测试数据，因此，这些数据有时也称 fixture，举例来说，Django 社区就使用 fixture 指启动应用时的初始数据。

不管怎样，本书中的 fixture 是 pytest 用于将测试前后进行预备、清理工作的代码分离出核心测试逻辑的一种机制。

pytest 正是因为有了 fixture 才表现得如此出色，许多测试人员放弃其他测试框架转向 pytest 都是因为 fixture。pytest 的 fixture 与 Django 的 fixture 不同，pytest 的 fixture 也不同于 unittest、nose 中的配置和销毁步骤。pytest 的 fixture 非常灵活，一旦你掌握了它，工作就会变得得心应手。但要熟练掌握它，离不开反复练习，现在就开始详细介绍 fixture。

3.1 通过 conftest.py 共享 fixture
Sharing Fixtures Through conftest.py

fixture 可以放在单独的测试文件里。如果你希望多个测试文件共享 fixture，可以在某个公共目录下新建一个 conftest.py 文件，将 fixture 放在其中。Tasks 项目的所有 fixture 都编写在 tasks_proj/tests/conftest.py 文件里，这样该项目中的所有测试都能共享其中的 fixture。

如果你希望 fixture 的作用域仅限于某个测试文件，那么将它写在该测试文件里。你也可以在 tests 目录的子目录下创建新的 conftest.py 文件，低一级的 conftest.py 中的 fixture 可以供所在目录及其子目录下的测试使用。Tasks 项目希望 fixture 共享给所有测试，因此它的 conftest.py 放在测试根目录 task_proj/tests 下最合适。

尽管 conftest.py 是 Python 模块，但它不能被测试文件导入。import conftest 的用法是不允许出现的。conftest.py 被 pytest 视作一个本地插件库（参考第 5 章）。可以把 tests/conftest.py 看成是一个供 tests 目录下所有测试使用的 fixture 仓库。

接下来修改针对 tasks_proj 的测试，以便使用 fixture。

3.2 使用 fixture 执行配置及销毁逻辑
Using Fixtures for Setup and Teardown

Tasks 项目中的大部分测试都假设 Tasks 数据库已经配置完成且正常启动了。我们有必要在测试完毕后进行清理，并断开与数据库的连接。其中大部分逻辑已经由 tasks 代码中的 tasks.start_tasks_db(<directory to store db>, 'tiny' or 'mongo') 和 tasks.stop_tasks_db() 实现，我们只需要调用即可。同时还需要一个临时目录。

pytest 中正好包含一个名为 tmpdir 的 fixture 供测试使用，无需担心清理的问题。这并非魔法，而是 pytest 团队的杰作。（tmpdir 及其近亲 tmpdir_factory 都将在第 4.1 节详细介绍。）

有了上面这些组件，tmpdir 使用起来会很方便。

```
ch3/a/tasks_proj/tests/conftest.py
import pytest
import tasks
from tasks import Task


@pytest.fixture()
def tasks_db(tmpdir):
    """Connect to db before tests, disconnect after."""
    # Setup : start db
```

```
tasks.start_tasks_db(str(tmpdir), 'tiny')

yield # this is where the testing happens

# Teardown : stop db
tasks.stop_tasks_db()
```

tmpdir 的值并非是一个字符串，而是一个目录对象，但由于指定了 __str__ 方法，所以它可以转换为字符串，并传递给 start_tasks_db()。我们仍然用 tiny 表示 TinyDB。

fixture 函数会在测试函数之前运行，但如果 fixture 函数包含 yield，那么系统会在 yield 处停止，转而运行测试函数，等测试函数执行完毕后再回到 fixture，继续执行 yield 后面的代码。因此，可以将 yield 之前的代码视为配置（setup）过程，将 yield 之后的代码视为清理（teardown）过程。无论测试过程中发生了什么，yield 之后的代码都会被执行。本例中没有让 yield 返回任何数据，读者可以自行尝试。

现在来修改针对 tasks.add() 的测试，以便使用 fixture。

ch3/a/tasks_proj/tests/func/test_add.py
```python
import pytest
import tasks
from tasks import Task

def test_add_returns_valid_id(tasks_db):
    """tasks.add(<valid task>) should return an integer."""
    # GIVEN an initialized tasks db
    # WHEN a new task is added
    # THEN returned task_id is of type int
    new_task = Task('do something')
    task_id = tasks.add(new_task)
    assert isinstance(task_id, int)
```

这里的主要修改是删除了文件中多余的 fixture，并将 tasks_db 添加到该测试的参数列表中。我习惯使用 GIVEN/WHEN/THEN 这样的注释来组织测试的结构，从而让模糊的逻辑变得更清晰。GIVEN an initialized tasks db 这句说明了为什么要使用 tasks_db 作为 fixture。

> **确保 Tasks 项目程序包安装完毕**
>
> 我们在第 2 章已学习了安装 Tasks 项目程序包，所以这里假定读者已安装完毕。如果你跳过了第 2 章，请先安装 tasks 包，安装命令：`cd code; pip install./tasks_proj/`。

3.3 使用--setup-show 回溯 fixture 的执行过程
Tracing Fixture Execution with –setup-show

如果直接运行上一节中的测试，则不会看到 fixture 的执行过程。

```
$ cd /path/to/code/
$ pip install ./tasks_proj/ # if not installed yet
$ cd /path/to/code/ch3/a/tasks_proj/tests/func
$ pytest -v test_add.py -k valid_id
===================== test session starts =====================
collected 3 items

test_add.py::test_add_returns_valid_id PASSED
===================== 2 tests deselected ======================
============= 1 passed, 2 deselected in 0.02 seconds ==========
```

我编写 fixture 时希望看到测试过程中执行的是什么，以及执行的先后顺序，pytest 提供的--setup-show 选项可以实现这个功能。

```
$ pytest --setup-show test_add.py -k valid_id
===================== test session starts =====================
collected 3 items
test_add.py
SETUP    S tmpdir_factory
    SETUP    F tmpdir (fixtures used: tmpdir_factory)
    SETUP    F tasks_db (fixtures used: tmpdir)
      func/test_add.py::test_add_returns_valid_id
        (fixtures used: tasks_db, tmpdir, tmpdir_factory).
    TEARDOWN F tasks_db
    TEARDOWN F tmpdir
TEARDOWN S tmpdir_factory
===================== 2 tests deselected ======================
============= 1 passed, 2 deselected in 0.02 seconds ==========
```

第3章 pytest Fixture

我们的测试被夹在正中间，pytest 将每一个 fixture 的执行分成 SETUP 和 TEARDOWN 两部分。从 test_add_returns_valid_id 向上看，可以发现 tmpdir 是运行在测试之前的，再往前是 tmpdir_factory。很显然，tmpdir 将 tmpdir_factory 作为它的 fixture。

fixture 名称前面的 F 和 S 代表的是 fixture 的作用范围，F 代表函数级别的作用范围，S 代表会话级别的作用范围（详见第 3.6 节）。

3.4 使用 fixture 传递测试数据
Using Fixtures for Test Data

fixture 非常适合存放测试数据，并且它可以返回任何数据。下面的例子返回了一个混合类型元素组成的 Python 元组。

```
ch3/test_fixtures.py
@pytest.fixture()
def a_tuple():
    """Return something more interesting."""
    return (1, 'foo', None, {'bar': 23})

def test_a_tuple(a_tuple):
    """Demo the a_tuple fixture."""
    assert a_tuple[3]['bar'] == 32
```

23 不等于 32，这将导致 test_a_tuple() 失败，于是可以观察到使用 fixture 的测试发生失败的情况。

```
$ cd /path/to/code/ch3
$ pytest test_fixtures.py::test_a_tuple
===================== test session starts =====================
collected 1 item

test_fixtures.py F
========================== FAILURES ===========================
_____ test_a_tuple _____
a_tuple = (1, 'foo', None, {'bar': 23})
    def test_a_tuple(a_tuple):
        """Demo the a_tuple fixture."""
>       assert a_tuple[3]['bar'] == 32
```

```
E           assert 23 == 32
test_fixtures.py:43: AssertionError
=================== 1 failed in 0.07 seconds ===================
```

除了堆栈跟踪的内容，pytest 还给出了具体引起 assert 异常的函数参数值。fixture 作为测试函数的参数，也会被堆栈跟踪并纳入测试报告。

假设 assert 异常（或任何类型的异常）就发生在 fixture，会怎么样呢？

```
$ pytest -v test_fixtures.py::test_other_data
===================== test session starts =====================
collected 1 item

test_fixtures.py::test_other_data ERROR
=========================== ERRORS ============================
_____ ERROR at setup of test_other_data _____
    @pytest.fixture()
    def some_other_data():
        """Raise an exception from fixture."""
        x = 43
>       assert x==42
E       assert 43==42
test_fixtures.py:24: AssertionError
=================== 1 error in 0.04 seconds ===================
```

首先，堆栈追踪正确定位了 fixture 函数中发生的 assert 异常；其次，test_other_data 并没有被报告为 FAIL，而是报告为 ERROR。这个区分很清楚，如果测试结果为 FAIL，用户就知道失败是发生在核心测试函数内，而不是发生在测试依赖的 fixture。

针对 Tasks 项目，我们会用一些数据相关的 fixture，例如返回值是一个包含多属性 task 的对象列表。

```
ch3/a/tasks_proj/tests/conftest.py
# Reminder of Task constructor interface
# Task(summary=None, owner=None, done=False, id=None)
# summary is required
# owner and done are optional
# id is set by database
@pytest.fixture()

def tasks_just_a_few():
    """All summaries and owners are unique."""
```

```python
    return (
        Task('Write some code', 'Brian', True),
        Task("Code review Brian's code", 'Katie', False),
        Task('Fix what Brian did', 'Michelle', False))

@pytest.fixture()
def tasks_mult_per_owner():
    """Several owners with several tasks each."""
    return (
        Task('Make a cookie', 'Raphael'),
        Task('Use an emoji', 'Raphael'),
        Task('Move to Berlin', 'Raphael'),

        Task('Create', 'Michelle'),
        Task('Inspire', 'Michelle'),
        Task('Encourage', 'Michelle'),

        Task('Do a handstand', 'Daniel'),
        Task('Write some books', 'Daniel'),
        Task('Eat ice cream', 'Daniel'))
```

这两个 fixture 可以直接用到测试里，也可以被其他 fixture 使用。接下来将用它们构建测试前的初始数据。

3.5 使用多个 fixture
Using Multiple Fixtures

我们已经知道 fixture tmpdir 可以使用 fixture tmpdir_factory，又在 fixture tasks_db 中使用了 fixture tmpdir。现在可以继续这样做，为测试数据的初始化生成更多的 fixture。

```python
# ch3/a/tasks_proj/tests/conftest.py
@pytest.fixture()
def db_with_3_tasks(tasks_db, tasks_just_a_few):
    """Connected db with 3 tasks, all unique."""
    for t in tasks_just_a_few:
        tasks.add(t)

@pytest.fixture()
def db_with_multi_per_owner(tasks_db, tasks_mult_per_owner):
    """Connected db with 9 tasks, 3 owners, all with 3 tasks."""
    for t in tasks_mult_per_owner:
        tasks.add(t)
```

两个 fixture 的形参中各包含两个 fixture（tasks_db 和返回数据集的 fixture）。数据集用于向数据库中添加 task，这样测试时使用的就不是空数据库了。

ch3/a/tasks_proj/tests/func/test_add.py
```python
def test_add_increases_count(db_with_3_tasks):
    """Test tasks.add() affect on tasks.count()."""
    # GIVEN a db with 3 tasks
    # WHEN another task is added
    tasks.add(Task('throw a party'))
    #  THEN the count increases by 1
    assert tasks.count() == 4
```

使用 fixture 的优势在于：用户在编写测试函数时可以只考虑核心测试逻辑，而不需要考虑测试前的准备工作。我习惯在注释中写 GIVEN/WHEN/THEN，并且总是在 fixture 中尽量多写 GIVEN。这样做有两个原因：一是增加测试的可读性和可维护性；二是 fixture 中的异常不会被报告为 FAIL，而是 ERROR。我不喜欢看到 test_add_increases_count() 的失败是由数据库初始化引起，这会令测试显得逻辑混乱，我希望 test_add_increases_count() 的失败只由 add() 引起。现在打开信息回溯选项，观察各层 fixture 的运行情况。

```
$ cd /path/to/code/ch3/a/tasks_proj/tests/func
$ pytest --setup-show test_add.py::test_add_increases_count
===================== test session starts ======================
collected 1 item

test_add.py
    SETUP    S tmpdir_factory
    SETUP    F tmpdir (fixtures used: tmpdir_factory)
    SETUP    F tasks_db (fixtures used: tmpdir)
    SETUP    F tasks_just_a_few
    SETUPF db_with_3_tasks (fixtures used: tasks_db, tasks_just_a_few)
      func/test_add.py::test_add_increases_count
        (fixtures used: db_with_3_tasks, tasks_db, tasks_just_a_few,
                  tmpdir, tmpdir_factory).
    TEARDOWN F db_with_3_tasks
    TEARDOWN F tasks_just_a_few
    TEARDOWN F tasks_db
    TEARDOWN F tmpdir
    TEARDOWN S tmpdir_factory
=================== 1 passed in 0.04 seconds ===================
```

这里又出现了 F 和 S，它们分别代表 fixture 的作用范围是函数级别和会话级别，下一节将重点讨论 fixture 的作用范围。

3.6 指定 fixture 作用范围
Specifying Fixture Scope

fixture 包含一个叫 scope（作用范围）的可选参数，用于控制 fixture 执行配置和销毁逻辑的频率。`@pytest.fixture()` 的 `scope` 参数有四个待选值：`function`、`class`、`module`、`session`（默认值为 `function`）。前面用到的 fixture 都没有指定 scope，因此它们的作用范围都是默认的函数级别。

以下是对各个 scope 的概述。

scope='function'

函数级别的 fixture 每个测试函数只需要运行一次。配置代码在测试用例运行之前运行，销毁代码在测试用例运行之后运行。function 是 scope 参数的默认值。

scope='class'

类级别的 fixture 每个测试类只需要运行一次，无论测试类里有多少类方法都可以共享这个 fixture。

scope='module'

模块级别的 fixture 每个模块只需要运行一次，无论模块里有多少个测试函数、类方法或其他 fixture 都可以共享这个 fixture。

scope='session'

会话级别的 fixture 每次会话只需要运行一次。一次 pytest 会话中的所有测试函数、方法都可以共享这个 fixture。

可以像这样使用作用范围。

ch3/test_scope.py

```python
"""Demo fixture scope."""

import pytest

@pytest.fixture(scope='function')
def func_scope():
    """A function scope fixture."""

@pytest.fixture(scope='module')
def mod_scope():
    """A module scope fixture."""

@pytest.fixture(scope='session')
def sess_scope():
    """A session scope fixture."""

@pytest.fixture(scope='class')
def class_scope():
    """A class scope fixture."""

def test_1(sess_scope, mod_scope, func_scope):
    """Test using session, module, and function scope fixtures."""

def test_2(sess_scope, mod_scope, func_scope):
    """Demo is more fun with multiple tests."""

@pytest.mark.usefixtures('class_scope')
class TestSomething():
    """Demo class scope fixtures."""
def test_3(self):
    """Test using a class scope fixture."""
def test_4(self):
    """Again, multiple tests are more fun."""
```

可以使用--setup-show命令行选项观察每个fixture被调用的次数，以及在各自作用范围下执行配置、销毁逻辑的顺序。

```
$ cd /path/to/code/ch3
$ pytest --setup-show test_scope.py
======================= test session starts ========================
collected 4 items

test_scope.py
SETUP    S sess_scope
  SETUP    M mod_scope
    SETUP    F func_scope
```

```
            test_scope.py::test_1
              (fixtures used: func_scope, mod_scope, sess_scope).
        TEARDOWN F func_scope
        SETUP    F func_scope
            test_scope.py::test_2
              (fixtures used: func_scope, mod_scope, sess_scope).
        TEARDOWN F func_scope
      SETUP    C class_scope
            test_scope.py::TestSomething::()::test_3 (fixtures used:
              class_scope).
            test_scope.py::TestSomething::()::test_4 (fixtures used:
              class_scope).
      TEARDOWN C class_scope
    TEARDOWN M mod_scope
TEARDOWN S sess_scope
===================== 4 passed in 0.01 seconds =====================
```

上面不仅出现了代表函数级别和会话级别的 F 和 S，还出现了代表类级别和模块级别的 C 和 M。

我们不难从代码中看出，作用范围虽然是由 fixture 自身定义的，但还是要强调 scope 参数是在定义 fixture 时定义的，而不是在调用 fixture 时定义的，因此，使用 fixture 的测试函数是无法改变 fixture 的作用范围的。

fixture 只能使用同级别的 fixture，或比自己级别更高的 fixture。比如，函数级别的 fixture 可以使用同级别的 fixture，也可以使用类级别、模块级别、会话级别的 fixture，但不能反过来。

修改 Tasks 项目的 fixture 作用范围
Changing Scope for Tasks Project Fixtures

学习了 fixture 的作用范围，现在来修改 Tasks 项目的 fixture 作用范围。

目前为止，每次测试都配置一个临时目录、新建一次数据库连接，这样显得有些浪费，实际上，只要保证每次测试时数据库有数据即可。

如果要将 tasks_db 的作用范围调整到会话级别，则应该选择 tmpdir_factory，因为 tmpdir 的作用范围是函数级别，而 tmpdir_factory 是会话级别。这里只需要改动一行代码(如果将 tmpdir 改为 tmpdir_factory 也算进去,就是两行)。

ch3/b/tasks_proj/tests/conftest.py
```python
import pytest
import tasks
from tasks import Task

@pytest.fixture(scope='session')
def tasks_db_session(tmpdir_factory):
    """Connect to db before tests, disconnect after."""
    temp_dir = tmpdir_factory.mktemp('temp')
    tasks.start_tasks_db(str(temp_dir), 'tiny')
    yield
    tasks.stop_tasks_db()

@pytest.fixture()
def tasks_db(tasks_db_session):
    """An empty tasks db."""
    tasks.delete_all()
```

改动 tasks_db,使其依赖于 tasks_db_session,并执行所有条目的删除指令。由于没有改动 tasks_db 的名字,所以所有依赖 tasks_db 的测试和 fixture 不需要做任何改动。

数据相关的 fixture 的作用只是返回值,没有必要一直运行,每次会话调用一次即可:

ch3/b/tasks_proj/tests/conftest.py
```python
# Reminder of Task constructor interface
# Task(summary=None, owner=None, done=False, id=None)
# summary is required
# owner and done are optional
# id is set by database

@pytest.fixture(scope='session')
def tasks_just_a_few():
    """All summaries and owners are unique."""
    return (
        Task('Write some code', 'Brian', True),
        Task("Code review Brian's code", 'Katie', False),
        Task('Fix what Brian did', 'Michelle', False))
```

```python
@pytest.fixture(scope='session')
def tasks_mult_per_owner():
    """Several owners with several tasks each."""
    return (
        Task('Make a cookie', 'Raphael'),
        Task('Use an emoji', 'Raphael'),
        Task('Move to Berlin', 'Raphael'),

        Task('Create', 'Michelle'),
        Task('Inspire', 'Michelle'),
        Task('Encourage', 'Michelle'),

        Task('Do a handstand', 'Daniel'),
        Task('Write some books', 'Daniel'),
        Task('Eat ice cream', 'Daniel'))
```

现在来看这些改动究竟起到了什么作用。

```
$ cd /path/to/code/ch3/b/tasks_proj
$ pytest
===================== test session starts =====================
collected 55 items

tests/func/test_add.py ...
tests/func/test_add_variety.py ............................
tests/func/test_add_variety2.py ............
tests/func/test_api_exceptions.py .......
tests/func/test_unique_id.py .
tests/unit/test_task.py ....
================== 55 passed in 0.17 seconds ==================
```

看起来一切顺利。现在对单个测试文件中的 fixture 进行信息回溯，观察各级别作用范围下的 fixture 是否都符合预期：

```
$ pytest --setup-show tests/func/test_add.py
======================== test session starts ========================
collected 3 items

tests/func/test_add.py
SETUP    S tmpdir_factory
SETUP    S tasks_db_session (fixtures used: tmpdir_factory)
      SETUP    F tasks_db (fixtures used: tasks_db_session)
        tests/func/test_add.py::test_add_returns_valid_id
          (fixtures used: tasks_db, tasks_db_session, tmpdir_factory).
      TEARDOWN F tasks_db
      SETUP    F tasks_db (fixtures used: tasks_db_session)
        tests/func/test_add.py::test_added_task_has_id_set
```

```
                (fixtures used: tasks_db, tasks_db_session, tmpdir_factory).
        TEARDOWN F tasks_db
          SETUP    F tasks_db (fixtures used: tasks_db_session)
SETUP     S tasks_just_a_few
          SETUP    F db_with_3_tasks (fixtures used: tasks_db,
tasks_just_a_few)
              tests/func/test_add.py::test_add_increases_count
                (fixtures used: db_with_3_tasks, tasks_db, tasks_db_session,
                                tasks_just_a_few, tmpdir_factory).
        TEARDOWN F db_with_3_tasks
        TEARDOWN F tasks_db
TEARDOWN  S tasks_just_a_few
TEARDOWN  S tasks_db_session
TEARDOWN  S tmpdir_factory
===================== 3 passed in 0.03 seconds =====================
===================== 3 passed in 0.03 seconds =====================
```

好的，看起来不错，tasks_db_session 每次会话中只调用一次，而现在更精炼的 tasks_db 只是在每次测试之前清理数据库。

3.7 使用 usefixtures 指定 fixture
Specifying Fixtures with usefixtures

目前为止用到 fixture 的测试，都是在测试函数的参数列表中指定 fixture，实际上，也可以用@pytest.mark.usefixtures('fixture1', 'fixture2')标记测试函数或类。使用 usefixtures，需要在参数列表中指定一个或多个 fixture 字符串。这对测试函数来讲其实意义不大，但非常适合测试类。

ch3/test_scope.py
```
@pytest.mark.usefixtures('class_scope')
class TestSomething():
    """Demo class scope fixtures."""

    def test_3(self):
    """Test using a class scope fixture."""

    def test_4(self):
    """Again, multiple tests are more fun."""
```

使用 usefixtures 和在测试方法中添加 fixture 参数，二者大体上是差不多的。区别之一在于只有后者才能够使用 fixture 的返回值。

3.8 为常用 fixture 添加 autouse 选项
Using autouse for Fixtures That Always Get Used

到目前为止，本章用到的 fixture 都是根据测试本身来命名的（或者针对示例的测试类使用 usefixtures）。我们可以通过指定 autouse=True 选项，使作用域内的测试函数都运行该 fixture，这与那些需要多次运行、但不依赖任何的系统状态或外部数据的代码配合得很好。以下是一个比较生硬的例子。

ch3/test_autouse.py

```python
"""Demonstrate autouse fixtures."""

import pytest
import time

@pytest.fixture(autouse=True, scope='session')
def footer_session_scope():
    """Report the time at the end of a session."""
    yield
    now = time.time()
    print('--')
    print('finished : {}'.format(time.strftime('%d %b %X',
time.localtime(now))))
    print('-----------------')

@pytest.fixture(autouse=True)
def footer_function_scope():
    """Report test durations after each function."""
    start = time.time()
    yield
    stop = time.time()
    delta = stop - start
    print('\ntest duration : {:0.3} seconds'.format(delta))

def test_1():
    """Simulate long-ish running test."""
    time.sleep(1)

def test_2():
    """Simulate slightly longer test."""
    time.sleep(1.23)
```

我们希望在每个测试中添加测试时间，在每段会话结束时添加日期和时间。以下是运行结果。

```
$ cd /path/to/code/ch3
$ pytest -v -s test_autouse.py
===================== test session starts =====================
collected 2 items

test_autouse.py::test_1 PASSED
test duration : 1.0 seconds
test_autouse.py::test_2 PASSED
test duration : 1.24 seconds
--
finished : 25 Jul 16:18:27
----------------
=================== 2 passed in 2.25 seconds ==================
```

虽然 autouse 特性不错，但它看起来更像一个特例，而非规则。我建议读者最好选择先前的 fixture 用法。

有些读者可能会问，为什么本章没有在 tasks_db 中使用 autouse。理由是在 Tasks 项目中使用 API 函数，必须在数据库初始化之前通过测试一探究竟，其间可能会抛出一些异常，如果强制对每个测试都进行良好的初始化，就有可能测试不到这些异常。

3.9 为 fixture 重命名
Renaming Fixtures

fixture 的名字展示在使用它的测试或其他 fixture 函数的参数列表上，通常会和 fixture 函数名保持一致，但 pytest 也允许使用@pytest.fixture()的 name 参数对 fixture 重命名。

```
ch3/test_rename_fixture.py
"""Demonstrate fixture renaming."""

import pytest

@pytest.fixture(name='lue')
def ultimate_answer_to_life_the_universe_and_everything():
    """Return ultimate answer."""
    return 42
```

```python
def test_everything(lue):
    """Use the shorter name."""
    assert lue == 42
```

这个 fixture 原来的名字是 fixture_with_a_name_much_longer_than_lue，重命名后变成了 lue。使用 --setup-show 选项能看到这个变化。

```
$ pytest --setup-show test_rename_fixture.py
======================= test session starts =======================
collected 1 items

test_rename_fixture.py
    SETUP    F lue
      test_rename_fixture.py::test_everything_2 (fixtures used: lue).
    TEARDOWN F lue
===================== 1 passed in 0.01 seconds ====================
```

如果你想找出 lue 是在哪里定义的，则可以为 pytest 指定 --fixtures 命令行选项，并提供所在测试文件名。pytest 将列举所有可供测试使用的 fixture，包括重命名的。

```
$ pytest --fixtures test_rename_fixture.py
======================= test session starts =======================
---------- fixtures defined from test_rename_fixture ----------
lue
    Return ultimate answer.
================ no tests ran in 0.01 seconds ================
```

由于输出内容比较多，所以这里做了省略。所幸我们自己重新定义的 fixture 在输出内容的底部，这是 pytest 预先设定好的，因此可以方便查询 lue 的定义。现在将该特性应用到 Tasks 项目上。

```
$ cd /path/to/code/ch3/b/tasks_proj
$ pytest --fixtures tests/func/test_add.py
=================== test session starts ===================
...
tmpdir_factory
    Return a TempdirFactory instance for the test session.
tmpdir
    Return a temporary directory path object which is
    unique to each test function invocation, created as
    a sub directory of the base temporary directory.
    The returned object is a `py.path.local`_path object.
------------------ fixtures defined from conftest ------------------
```

```
    tasks_db_session
        Connect to db before tests, disconnect after.
    tasks_db
        An empty tasks db.
    tasks_just_a_few
        All summaries and owners are unique.
    tasks_mult_per_owner
        Several owners with several tasks each.
    db_with_3_tasks
        Connected db with 3 tasks, all unique.
    db_with_multi_per_owner
        Connected db with 9 tasks, 3 owners, all with 3 tasks.
=================== no tests ran in 0.01 seconds ====================
```

很好，conftest.py 中的所有 fixture 都显示出来了。pytest 内建 fixture 列表的末尾是 tmpdir 和 tmpdir_factory，之前我们用过。

3.10　Fixture 的参数化
Parametrizing Fixtures

第 2.8 节已经介绍过测试的参数化，也可以对 fixture 做参数化处理。为了和先前的例子对应，我们仍然使用 task 对象列表、task 标识列表和比较函数。

```
ch3/b/tasks_proj/tests/func/test_add_variety2.py
import pytest
import tasks
from tasks import Task
tasks_to_try = (Task('sleep', done=True),
                Task('wake', 'brian'),
                Task('breathe', 'BRIAN', True),
                Task('exercise', 'BrIaN', False))

task_ids = ['Task({},{},{})'.format(t.summary, t.owner, t.done)
            for t in tasks_to_try]

def equivalent(t1, t2):
    """Check two tasks for equivalence."""
    return ((t1.summary == t2.summary) and
            (t1.owner == t2.owner) and
            (t1.done == t2.done))
```

现在对一个名为 a_task 的 fixture 进行参数化处理。

```
ch3/b/tasks_proj/tests/func/test_add_variety2.py
```
```python
@pytest.fixture(params=tasks_to_try)
def a_task(request):
    """Using no ids."""
    return request.param

def test_add_a(tasks_db, a_task):
    """Using a_task fixture (no ids)."""
    task_id = tasks.add(a_task)
    t_from_db = tasks.get(task_id)
    assert equivalent(t_from_db, a_task)
```

fixture 参数列表中的 request 也是 pytest 内建的 fixture 之一，代表 fixture 的调用状态，第 4 章将详细介绍。它有一个 param 字段，会被 @pytest.fixture(params= tasks_to_try)的 params 列表中的一个元素填充。

a_task 逻辑非常简单，仅以 request.param 作为返回值供测试使用。因为 task 对象列表包含四个 task 对象，所以 a_task 将被测试调用四次。

```
$ cd /path/to/code/ch3/b/tasks_proj/tests/func
$ pytest -v test_add_variety2.py::test_add_a
===================== test session starts =====================
collected 4 items

test_add_variety2.py::test_add_a[a_task0] PASSED
test_add_variety2.py::test_add_a[a_task1] PASSED
test_add_variety2.py::test_add_a[a_task2] PASSED
test_add_variety2.py::test_add_a[a_task3] PASSED
==================== 4 passed in 0.03 seconds ====================
```

由于未指定 id，pytest 用 fixture 名加一串数字作为 task 标识。就像参数化测试一样，现在以相同的字符串列表为其指定 id。

```
ch3/b/tasks_proj/tests/func/test_add_variety2.py
```
```python
@pytest.fixture(params=tasks_to_try, ids=task_ids)
def b_task(request):
    """Using a list of ids."""
    return request.param

def test_add_b(tasks_db, b_task):
    """Using b_task fixture, with ids."""
    task_id = tasks.add(b_task)
    t_from_db = tasks.get(task_id)
    assert equivalent(t_from_db, b_task)
```

给 task 指定更友好的标识。

```
$ pytest -v test_add_variety2.py::test_add_b
===================== test session starts =====================
collected 4 items

test_add_variety2.py::test_add_b[Task(sleep,None,True)] PASSED
test_add_variety2.py::test_add_b[Task(wake,brian,False)] PASSED
test_add_variety2.py::test_add_b[Task(breathe,BRIAN,True)] PASSED
test_add_variety2.py::test_add_b[Task(exercise,BrIaN,False)] PASSED
=================== 4 passed in 0.04 seconds ==================
```

ids 参数也可以被指定为一个函数（前面是列表），供 pytest 生成 task 标识。

ch3/b/tasks_proj/tests/func/test_add_variety2.py
```
def id_func(fixture_value):
    """A function for generating ids."""
    t = fixture_value
    return 'Task({},{},{})'.format(t.summary, t.owner, t.done)

@pytest.fixture(params=tasks_to_try, ids=id_func)
def c_task(request):
    """Using a function (id_func) to generate ids."""
    return request.param

def test_add_c(tasks_db, c_task):
    """Use fixture with generated ids."""
    task_id = tasks.add(c_task)
    t_from_db = tasks.get(task_id)
    assert equivalent(t_from_db, c_task)
```

该函数将作用于 params 列表中的每一个元素。params 参数是一个 Task 对象列表，id_func() 将调用单个 Task 对象，这让我们能够使用 namedtuple 访问器方法访问单个 Task 对象，以便为每个 Task 对象生成标识符。这看起来会比之前使用列表推导式提前生成完整对象列表的方案简洁一些，但结果是一致的。

```
$ pytest -v test_add_variety2.py::test_add_c
===================== test session starts =====================
collected 4 items

test_add_variety2.py::test_add_c[Task(sleep,None,True)] PASSED
```

```
test_add_variety2.py::test_add_c[Task(wake,brian,False)] PASSED
test_add_variety2.py::test_add_c[Task(breathe,BRIAN,True)] PASSED
test_add_variety2.py::test_add_c[Task(exercise,BrIaN,False)] PASSED
==================== 4 passed in 0.04 seconds ====================
```

对测试函数进行参数化处理，可以多次运行的只是该测试函数，而使用参数化 fixture，每个使用该 fixture 的测试函数都可以被运行多次。fixture 的这一特性非常强大。

3.11 参数化 Tasks 项目中的 fixture
Parametrizing Fixtures in the Tasks Project

现在来看看如何在 Tasks 项目中应用参数化的 fixture。到目前为止，我们的测试使用的数据库都是 TinyDB，现在我们希望测试既可以使用 TinyDB，也可以使用 MongoDB。

具体使用哪个数据库是由 `tasks_db_session` fixture 中调用的 `start_tasks_db()` 函数决定的。

```python
ch3/b/tasks_proj/tests/conftest.py
import pytest
import tasks
from tasks import Task


@pytest.fixture(scope='session')
def tasks_db_session(tmpdir_factory):
    """Connect to db before tests, disconnect after."""
    temp_dir = tmpdir_factory.mktemp('temp')
    tasks.start_tasks_db(str(temp_dir), 'tiny')
    yield
    tasks.stop_tasks_db()


@pytest.fixture()
def tasks_db(tasks_db_session):
    """An empty tasks db."""
    tasks.delete_all()
```

`start_tasks_db()` 函数调用 `db_type` 参数，然后切换到负责相应数据库交互的子系统。

tasks_proj/src/tasks/api.py
```
def start_tasks_db(db_path, db_type): # type: (str, str) -> None
    """Connect API functions to a db."""
    if not isinstance(db_path, string_types):
        raise TypeError('db_path must be a string')
    global _tasksdb
    if db_type == 'tiny':
        import tasks.tasksdb_tinydb
        _tasksdb = tasks.tasksdb_tinydb.start_tasks_db(db_path)
    elif db_type == 'mongo':
        import tasks.tasksdb_pymongo
        _tasksdb = tasks.tasksdb_pymongo.start_tasks_db(db_path)
    else:
        raise ValueError("db_type must be a 'tiny' or 'mongo'")
```

为了使用 MongoDB，需要将所有测试的 db_type 都指定为 mongo。

ch3/c/tasks_proj/tests/conftest.py
```
import pytest❶
import tasks❶
from tasks import Task

#@pytest.fixture(scope='session', params=['tiny',])
@pytest.fixture(scope='session', params=['tiny', 'mongo'])
def tasks_db_session(tmpdir_factory, request):
    """Connect to db before tests, disconnect after."""
    temp_dir = tmpdir_factory.mktemp('temp')
    tasks.start_tasks_db(str(temp_dir), request.param)
    yield # this is where the testing happens
    tasks.stop_tasks_db()

@pytest.fixture()❷
def tasks_db(tasks_db_session):
    """An empty tasks db."""
    tasks.delete_all()
```

我在 fixture 装饰器中添加了 params=['tiny', 'mongo']，并将 request 添加到 temp_db 参数列表中，同时我将 db_type 的值设定为了 request.param，避免直接使用'tiny'或'mongo'。

运行参数化测试或参数化 fixture 时，如果命令行选项中指定了--verbose 或-v，pytest 则会根据参数值为每个测试用例分别做标记。这些值本身都是字符串类型，可以直接使用。

> **安装 MongoDB**
>
>
> 上面的示例中用到了 MongoDB，请先安装 MongoDB 和 pymongo。我一直在使用 MongoDB 的社区版本做测试（https://www.mongodb.com/download-center）。pymongo 可以通过 `pip install pymongo` 安装。本书用到 MongoDB 的地方不多，除了这里，还有一处是第 7 章的 debugger 的例子。

以下是运行情况。

```
$ cd /path/to/code/ch3/c/tasks_proj
$ pip install pymongo
$ pytest -v --tb=no
====================== test session starts ======================
collected 92 items

test_add.py::test_add_returns_valid_id[tiny] PASSED
test_add.py::test_added_task_has_id_set[tiny] PASSED
test_add.py::test_add_increases_count[tiny] PASSED
test_add_variety.py::test_add_1[tiny] PASSED
test_add_variety.py::test_add_2[tiny-task0] PASSED
test_add_variety.py::test_add_2[tiny-task1] PASSED
...
test_add.py::test_add_returns_valid_id[mongo] FAILED
test_add.py::test_added_task_has_id_set[mongo] FAILED
test_add.py::test_add_increases_count[mongo] PASSED
test_add_variety.py::test_add_1[mongo] FAILED
test_add_variety.py::test_add_2[mongo-task0] FAILED
...
============= 42 failed, 50 passed in 4.94 seconds =============
```

看起来不太妙，在使用这个 MongoDB 版本之前，还需要做一些 debug 的工作。第 7.1 节将介绍如何 debug，在那之前，还是继续使用 TinyDB。

3.12 练习
Exercises

1. 创建一个测试文件 `test_fixtures.py`。

2. 使用 `@pytest.fixture()` 装饰器编写几个用于返回数据的 fixture 函数，例如可以返回列表、字典、元组等。

3. 为每个 fixture 至少编写一个测试函数。

4. 编写一对使用相同 fixture 的测试。

5. 执行 `pytest --setup-show test_fixtures.py`，检查所有的 fixture 是不是都在测试之前运行。

6. 为练习 4 中的 fixture 添加 `scope='module'`。

7. 重新运行 `pytest --setup-show test_fixtures.py`，看看发生了哪些变化？

8. 修改练习 6 中的 fixture，将 `return<data>` 替换为 `yield <data>`。

9. 在 yield 前后都添加 print 语句。

10. 执行 `pytest -s -v test_fixtures.py`，从输出信息中你看出了什么？

3.13 预告
What's Next

pytest 的 fixture 非常灵活，既可以作为测试前后配置及销毁阶段的公共组件，也可以用于切换到不同的子系统（比如从 TinyDB 到 MongoDB）。我自己在测试工作中大量使用 fixture，希望大家也学会使用它。

读者在本章学习了编写自己的 fixture，也使用了 pytest 内置的 fixture（`tmpdir` 和 `tmpdir_factory`），第 4 章将详细介绍 pytest 内置的 fixture。

第 4 章

内置 Fixture
Builtin Fixtures

第 3 章介绍了什么是 fixture，如何通过编写 fixture 为测试提供数据，如何执行测试前的配置逻辑和测试后的销毁逻辑。我们还使用了 conftest.py 文件在多个测试文件中共享 fixture。第 3 章的 Tasks 项目用到了以下 fixture：tasks_db_session、tasks_just_a_few、tasks_mult_per_owner、tasks_db、db_with_3_tasks 和 db_ with_multi_per_owner。它们都定义在 conftest.py 文件里，Tasks 项目的所有测试函数都可以调用。

常见的 fixture 都支持复用，pytest 内置了一些常用的 fixture。我们已经在第 3 章见到了 tmpdir 和 tmpdir_factory。本章会更详细地介绍它们。

pytest 内置的 fixture 可以大幅简化测试工作。例如在处理临时文件时，pytest 内置的 fixture 可以识别命令行参数、在多个测试会话间通信、校验输出流、更改环境变量、审查错误警报，等等。内置 fixture 是对 pytest 核心功能的扩展。下面介绍几个常见 fixture 的用法。

4.1 使用 tmpdir 和 tmpdir_factory
Using tmpdir and tmpdir_factory

内置的 tmpdir 和 tmpdir_factory 负责在测试开始运行前创建临时文件目录，并在测试结束后删除。在 Tasks 项目里，我们需要一个目录来存放 MongoDB 和 TinyDB 使用的临时数据文件。我们希望测试时使用的是临时数据库，并且数据在测试会话结束后会销毁，所以使用 tmpdir 和 tmpdir_factory 来创建和删除临时目录。

如果测试代码要对文件进行读/写操作，那可以使用 tmpdir 或 tmpdir_factory 来创建文件或目录。单个测试使用 tmpdir，多个测试使用 tmpdir_factory。

tmpdir 的作用范围是函数级别，tmpdir_factory 的作用范围是会话级别。单个测试需要临时目录或文件应该使用 tmpdir。如果每个测试函数都要重新创建目录或文件，也请使用 tmpdir。下面是使用 tmpdir 的一个例子。

```
ch4/test_tmpdir.py
def test_tmpdir(tmpdir):
    # tmpdir already has a path name associated with it
    # join() extends the path to include a filename
    # the file is created when it's written to
    a_file = tmpdir.join('something.txt')

    # you can create directories
    a_sub_dir = tmpdir.mkdir('anything')

    # you can create files in directories (created when written)
    another_file = a_sub_dir.join('something_else.txt')

    # this write creates 'something.txt'
    a_file.write('contents may settle during shipping')

    # this write creates 'anything/something_else.txt'
    another_file.write('something different')

    # you can read the files as well
    assert a_file.read() == 'contents may settle during shipping'
    assert another_file.read() == 'something different'
```

4.1 使用 tmpdir 和 tmpdir_factory

tmpdir 的返回值是 py.path.local 类型的一个对象[1]。这看起来满足了我们对临时目录和文件的要求。但要注意，因为 tmpdir 的作用范围是函数级别的，所以只能针对测试函数使用 tmpdir 创建文件或目录。如果需要 fixture 作用范围高于函数级别（如类、模块、会话级别），则应使用 tmpdir_factory。

tmpdir_factory 与 tmpdir 很像，但它们有不同的接口。我们在第 3.6 节讨论过，作用范围是函数级别的 fixture，每个测试函数只需运行一次；模块级别的 fixture，每个测试模块只需运行一次；类级别的 fixture，每个测试类只需运行一次；会话级别的 fixture，每个会话只需运行一次。因此，在会话级别的 fixture 中，创建的资源可以在整个会话期间使用。

为了展示 tmpdir 和 tmpdir_factory 的相似性，我修改了 tmpdir 的例子，使用 tmpdir_factory 替换它。

```
ch4/test_tmpdir.py
def test_tmpdir_factory(tmpdir_factory):
    # you should start with making a directory
    # a_dir acts like the object returned from the tmpdir fixture
    a_dir = tmpdir_factory.mktemp('mydir')

    # base_temp will be the parent dir of 'mydir'
    # you don't have to use getbasetemp()
    # using it here just to show that it's available
    base_temp = tmpdir_factory.getbasetemp()
    print('base:', base_temp)

    # the rest of this test looks the same as the 'test_tmpdir()'
    # example except I'm using a_dir instead of tmpdir
    a_file = a_dir.join('something.txt')
    a_sub_dir = a_dir.mkdir('anything')
    another_file = a_sub_dir.join('something_else.txt')

    a_file.write('contents may settle during shipping')
    another_file.write('something different')
    assert a_file.read() == 'contents may settle during shipping'
    assert another_file.read() == 'something different'
```

---

[1] http://py.readthedocs.io/en/latest/path.html

我使用 tmpdir_factory.mktemp('mydir')创建了一个名为 a_dir 的目录，在此函数内，它的效果与使用 tmpdir 的效果是一样的。

getbasetemp()函数返回了该会话使用的根目录。print 语句用来展示系统中该目录的位置。让我们来看看。

```
$ cd /path/to/code/ch4
$ pytest -q -s test_tmpdir.py::test_tmpdir_factory
base: /private/var/folders/53/zv4j_zc506x2xq25l31qxvxm0000gn\
    /T/pytest-of-okken/pytest-732
.
1 passed in 0.04 seconds
```

pytest-NUM 会随着会话的递增而递增。pytest 会记录最近几次会话使用的根目录，更早的根目录记录则会被清理掉。如果你想在测试后检查文件，那么了解这一点会很有用。

你也可以使用 pytest –basetemp=mydir 指定自己的根目录。

在其他作用范围内使用临时目录
Using Temporary Directories for Other Scopes

tmpdir_factory 的作用范围是会话级别的，tmpdir 的作用范围是函数级别的。如果需要模块或类级别作用范围的目录，该怎么办？这时，可以利用 tmpdir_factory 再创建一个 fixture。

假定有一个测试模块，其中有很多测试用例要读取一个 JSON 文件。我们可以在模块本身或 conftest.py 中创建一个作用范围是模块级别的 fixture，用于配置该文件，如下例。

ch4/authors/conftest.py
```python
"""Demonstrate tmpdir_factory."""

import json
import pytest


@pytest.fixture(scope='module')
def author_file_json(tmpdir_factory):
```

```python
"""Write some authors to a data file."""
python_author_data = {
    'Ned': {'City': 'Boston'},
    'Brian': {'City': 'Portland'},
    'Luciano': {'City': 'Sau Paulo'}
}

file = tmpdir_factory.mktemp('data').join('author_file.json')
print('file:{}'.format(str(file)))

with file.open('w') as f:
    json.dump(python_author_data, f)
return file
```

author_file_json()这个新 fixture 创建了一个临时目录 data，并在此目录下创建了一个文件 author_file.json。然后它将 python_author_data 字典数据写入 JSON 文件中。因为这个新 fixture 的作用范围是模块级别的，所以该 JSON 文件只需要被每个模块创建一次。

ch4/authors/test_authors.py
```python
"""Some tests that use temp data files."""
import json

def test_brian_in_portland(author_file_json):
    """A test that uses a data file."""
    with author_file_json.open() as f:
        authors = json.load(f)
    assert authors['Brian']['City'] == 'Portland'

def test_all_have_cities(author_file_json):
    """Same file is used for both tests."""
    with author_file_json.open() as f:
        authors = json.load(f)
    for a in authors:
        assert len(authors[a]['City']) > 0
```

这里两个测试用例将使用同一个 JSON 文件。如果一个测试数据文件能服务于多个测试用例，那么就没必要为每个用例创建一个数据文件。

4.2 使用 pytestconfig
Using pytestconfig

内置的 pytestconfig 可以通过命令行参数、选项、配置文件、插件、运行目录等方式来控制 pytest。pytestconfig 是 request.config 的快捷方式，它在 pytest 文档里有时候被称为 "pytest 配置对象"。

要理解 pytestconfig 是如何工作的，可以查看如何添加一个自定义的命令行选项，然后在测试用例中读取该选项。你可以直接从 pytestconfig 里读取自定义的命令行选项，但是，为了让 pytest 能够解析它，还需要使用 hook 函数。hook 函数（参见第 5 章的介绍）是另一种控制 pytest 的方法，在插件中频繁使用。添加自定义的命令行选项，然后从 pytestconfig 中读取是很常见的操作。

下面使用 pytest 的 hook 函数 pytest_addoption 添加几个命令行选项。

```
ch4/pytestconfig/conftest.py
def pytest_addoption(parser):
    parser.addoption("--myopt", action="store_true",
                     help="some boolean option")
    parser.addoption("--foo", action="store", default="bar",
                     help="foo: bar or baz")
```

以 pytest_addoption 添加的命令行选项必须通过插件来实现，或者在项目顶层目录的 conftest.py 文件中完成。它所在的 conftest.py 不能处于测试子目录下。

上面的代码添加了新选项--myopt 和--foo<value>，还更改了帮助信息。

```
$ pytest --help
usage: pytest
[options] [file_or_dir] [file_or_dir] [...]
...
custom options:
  --myopt              some boolean option

  --foo=FOO            foo: bar or baz
...
```

接下来就可以在测试用例中使用这些选项了。

ch4/pytestconfig/test_config.py
```python
import pytest

def test_option(pytestconfig):
    print('"foo" set to:', pytestconfig.getoption('foo'))
    print('"myopt" set to:', pytestconfig.getoption('myopt'))
```

让我们看看它是如何工作的。

```
$ pytest -s -q test_config.py::test_option
"foo" set to: bar
"myopt" set to: False
.1
passed in 0.01 seconds
$ pytest -s -q --myopt test_config.py::test_option
"foo" set to: bar
"myopt" set to: True
.1
passed in 0.01 seconds
$ pytest -s -q --myopt --foo baz test_config.py::test_option
"foo" set to: baz
"myopt" set to: True
.1
passed in 0.01 seconds
```

因为 pytestconfig 是一个 fixture，所以它也可以被其他 fixture 使用。如果你喜欢，也可以为这些选项创建 fixture。

ch4/pytestconfig/test_config.py
```python
@pytest.fixture()
def foo(pytestconfig):
    return pytestconfig.option.foo

@pytest.fixture()
def myopt(pytestconfig):
    return pytestconfig.option.myopt

def test_fixtures_for_options(foo, myopt):
    print('"foo" set to:', foo)
    print('"myopt" set to:', myopt)
```

你也可以使用内置的选项（而不仅仅是你添加的选项），以及那些 pytest 启动时的信息（目录、参数等）。

下面是一个有若干配置信息和选项的例子。

```
ch4/pytestconfig/test_config.py
def test_pytestconfig(pytestconfig):
    print('args :', pytestconfig.args)
    print('inifile :', pytestconfig.inifile)
    print('invocation_dir :', pytestconfig.invocation_dir)
    print('rootdir :', pytestconfig.rootdir)
    print('-k EXPRESSION :', pytestconfig.getoption('keyword'))
    print('-v, --verbose :', pytestconfig.getoption('verbose'))
    print('-q, --quiet :', pytestconfig.getoption('quiet'))
    print('-l, --showlocals:', pytestconfig.getoption('showlocals'))
    print('--tb=style :', pytestconfig.getoption('tbstyle'))
```

第 6 章介绍 ini 文件时还会再次用到 pytestconfig。

4.3 使用 cache
Using cache

通常我们认为每个测试用例都是相互独立的，因此需要保证测试结果不依赖于测试顺序，即以不同的顺序运行测试用例，得到相同的测试结果。同时，我们也希望每段测试会话可以重复，而不会因为上一段会话的运行影响这一段的测试行为。

然而，有时从一段测试会话传递信息给下一段会话很有用。这时，可以使用 pytest 内置的 cache。

cache 的作用是存储一段测试会话的信息，在下一段测试会话中使用。使用 pytest 内置的 --last-failed 和 --failed-first 标识可以很好地展示 cache 的功能。让我们看看 cache 是如何存储这些标识数据的。

下面是 --last-failed 和 --failed-first 的帮助信息，包括一些 cache 选项。

```
$ pytest --help
...
  --lf, --last-failed    rerun only the tests that failed at the last run(or
                         all if none failed)
  --ff, --failed-first   run all tests but run the last failures first.
                         This may re-order tests and thus lead to repeated
                         fixture setup/teardown
  --cache-show           show cache contents, don't perform collection or tests
  --cache-clear          remove all cache contents at start of test run.
...
```

为了展示它们的运作方式，我们使用下面两个测试用例。

ch4/cache/test_pass_fail.py
```python
def test_this_passes():
    assert 1 == 1


def test_this_fails():
    assert 1 == 2
```

我们使用--verbose 标识来显示函数名字，使用--tb=no 标识来隐藏堆栈信息。

```
$ cd /path/to/code/ch4/cache
$ pytest --verbose --tb=no test_pass_fail.py
==================== test session starts ====================
collected 2 items

test_pass_fail.py::test_this_passes PASSED
test_pass_fail.py::test_this_fails FAILED
============ 1 failed, 1 passed in 0.05 seconds =============
```

如果再次运行并使用--ff 或者--failed-first 标识，那么之前运行未通过的测试用例会首先运行，然后才是其他用例。

```
$ pytest --verbose --tb=no --ff test_pass_fail.py
==================== test session starts ====================
run-last-failure: rerun last 1 failures first
collected 2 items

test_pass_fail.py::test_this_fails FAILED
test_pass_fail.py::test_this_passes PASSED
============ 1 failed, 1 passed in 0.04 seconds =============
```

也可以使用--lf 或--last-failed 仅运行上次未通过的测试用例。

```
$ pytest --verbose --tb=no --lf test_pass_fail.py
=================== test session starts ====================
run-last-failure: rerun last 1 failures
collected 2 items

test_pass_fail.py::test_this_fails FAILED
=================== 1 tests deselected =====================
========== 1 failed, 1 deselected in 0.05 seconds ==========
```

在讲解失败数据是如何存储并被使用前, 让我们看另外一个例子, 它让 --lf 和 --ff 标识的作用变得更明显。

下面是一个参数化的测试, 其中一个用例未通过。

ch4/cache/test_few_failures.py

```python
"""Demonstrate -lf and -ff with failing tests."""

import pytest
from pytest import approx

testdata = [
    # x, y, expected
    (1.01, 2.01, 3.02),
    (1e25, 1e23, 1.1e25),
    (1.23, 3.21, 4.44),
    (0.1, 0.2, 0.3),
    (1e25, 1e24, 1.1e25)
]


@pytest.mark.parametrize("x,y,expected", testdata)
def test_a(x, y, expected):
    """Demo approx()."""
    sum_ = x + y
    assert sum_ == approx(expected)
```

输出如下。

```
$ cd /path/to/code/ch4/cache
$ pytest -q test_few_failures.py
.F...
====================== FAILURES ========================
_____ test_a[1e+25-1e+23-1.1e+25] _____
x = 1e+25, y = 1e+23, expected = 1.1e+25

    @pytest.mark.parametrize("x,y,expected", testdata)
    def test_a(x,y,expected):
        sum_ = x + y
```

```
>       assert sum_ == approx(expected)
E       assert 1.01e+25 == 1.1e+25 ± 1.1e+19
E        + where 1.1e+25 ± 1.1e+19 = approx(1.1e+25)

test_few_failures.py:17: AssertionError
1 failed, 4 passed in 0.06 seconds
```

也许你一眼就看出了错误，但让我们假装测试很长很复杂，错误不是那么明显。现在为了让失败再次出现，请单独运行一遍这个测试用例（可以在命令行中指定）。

```
$ pytest -q "test_few_failures.py::test_a[1e+25-1e+23-1.1e+25]"
```

如果不想复制/粘贴，或者希望重新运行多个失败的测试用例，则使用 `--lf` 更容易。调试失败的测试用例时，另外一个好用的选项是 `--showlocals`（简写为 `-l`）。

```
$ pytest -q --lf -l test_few_failures.py
F
===================== FAILURES =====================
_____ test_a[1e+25-1e+23-1.1e+25] _____
x = 1e+25, y = 1e+23, expected = 1.1e+25

    @pytest.mark.parametrize("x,y,expected", testdata)
    def test_a(x,y,expected):
        sum_ = x + y
>       assert sum_ == approx(expected)
E       assert 1.01e+25 == 1.1e+25 ± 1.1e+19
E        + where 1.1e+25 ± 1.1e+19 = approx(1.1e+25)

expected   = 1.1e+25
sum_       = 1.01e+25
x          = 1e+25
y          = 1e+23

test_few_failures.py:17: AssertionError
================= 4 tests deselected =================
1 failed, 4 deselected in 0.05 seconds
```

失败的原因现在更明显了。

为了记住上次测试失败的用例，pytest 存储了上个测试会话中测试失败的信息。你可以使用 `--cache-show` 标识来显示存储的信息。

```
$ pytest --cache-show
===================== test session starts =====================
------------------------ cache values ------------------------
cache/lastfailed contains:
  {'test_few_failures.py::test_a[1e+25-1e+23-1.1e+25]': True}
================= no tests ran in 0.00 seconds =================
```

还可以查看 cache 目录。

```
$ cat .cache/v/cache/lastfailed
{
  "test_few_failures.py::test_a[1e+25-1e+23-1.1e+25]": true
}
```

你也可以在测试会话开始前传入 --clear-cache 标识来清空缓存。

cache 的功能远不止提供 --lf 和 --ff 标识。我们来创建一个 fixture，记录测试的耗时，并存储到 cache 里，如果接下来的测试耗时大于之前的两倍，就抛出超时异常。

cache 的接口很简单。

```
cache.get(key, default)
cache.set(key, value)
```

习惯上，键名以应用名字或插件名字开始，接着是/，然后是分隔开的键字符串。键值可以是任何可转化成 JSON 的东西，因为在 .cache 目录里是用 JSON 格式存储的。

下面是用于测试计时的 fixture。

```
ch4/cache/test_slower.py
@pytest.fixture(autouse=True)
def check_duration(request, cache):
    key = 'duration/' + request.node.nodeid.replace(':', '_')
    # nodeid's can have colons
    # keys become filenames within .cache
    # replace colons with something filename safe
    start_time = datetime.datetime.now()
    yield
    stop_time = datetime.datetime.now()
    this_duration = (stop_time - start_time).total_seconds()
    last_duration = cache.get(key, None)
```

```
        cache.set(key, this_duration)
    if last_duration is not None:
        errorstring = "test duration over 2x last duration"
        assert this_duration <= last_duration * 2, errorstring
```

因为 fixture 设置为了 autouse，所以它不需要被测试用例引用。request 对象用来抓取键名中的 nodeid。nodeid 是一个独特的标识，即便是在参数化测试中也能使用。我们假设以'duration/'字符串开始的 cache 键名是合法的。yield 语句前面的代码在测试函数之前运行，yield 语句后的代码在测试函数结束后运行。

现在需要一些运行时间各不相同的测试用例。

ch4/cache/test_slower.py
```python
@pytest.mark.parametrize('i', range(5))
def test_slow_stuff(i):
    time.sleep(random.random())
```

你可能不想为此编写大量的测试用例，我用 random 函数和动态参数生成了一批测试用例，它们可随机休眠一段时间，时间都小于 1 秒。下面来运行几次。

```
$ cd /path/to/code/ch4/cache
$ pytest -q --cache-clear test_slower.py
.....
5 passed in 2.10 seconds
$ pytest -q --tb=line test_slower.py
.E..E.E.
============================= ERRORS =============================
_____ ERROR at teardown of test_slow_stuff[0] _____
E AssertionError: test duration over 2x last duration
assert 0.954312 <= (0.380536 * 2)
_____ ERROR at teardown of test_slow_stuff[2] _____
E AssertionError: test duration over 2x last duration
assert 0.821745 <= (0.152405 * 2)
_____ ERROR at teardown of test_slow_stuff[3] _____
E AssertionError: test duration over 2x last duration
assert 1.001032 <= (0.36674 * 2)
5 passed, 3 error in 3.83 seconds
```

有意思，让我们看看 cache 里有什么。

```
$ pytest -q --cache-show
------------------------ cache values ------------------------
cache/lastfailed contains:
  {'test_slower.py::test_slow_stuff[0]': True,
   'test_slower.py::test_slow_stuff[2]': True,
   'test_slower.py::test_slow_stuff[3]': True}
duration/test_slower.py__test_slow_stuff[0] contains:
  0.954312
duration/test_slower.py__test_slow_stuff[1] contains:
  0.915539
duration/test_slower.py__test_slow_stuff[2] contains:
  0.821745
duration/test_slower.py__test_slow_stuff[3] contains:
  1.001032
duration/test_slower.py__test_slow_stuff[4] contains:
  0.031884

no tests ran in 0.01 seconds
```

因为 cache 数据有前缀，所以不难发现 duration 数据。lastfailed 功能作用在一条 cache 记录上是很有趣的。duration 数据记录在每个测试用例对应的 cache 里。让我们按照 lastfailed 模式写入自己的数据。

接下来的每个测试都将读/写 cache。可以把原先的 fixture 拆分为两个小 fixture：一个作用范围是函数级别，用于测量运行时间；另一个作用范围是会话级别，用来读/写 cache。可如果这样做，就不能使用 cache fixture 了，因为它的作用范围是函数级别的。所幸在查看 GitHub 上 cache 的代码后[1]，我们发现 cache 只返回了 request.config.cache，而它适用于各个级别的作用范围。

下面是重构后的代码。

ch4/cache/test_slower_2.py
```
Duration = namedtuple('Duration', ['current', 'Last'])

@pytest.fixture(scope='session')
def duration_cache(request):
    key = 'duration/testdurations'
    d = Duration({}, request.config.cache.get(key, {}))
    yield d
```

---

[1] https://github.com/pytest-dev/pytest/blob/master/_pytest/cacheprovider.py

```
        request.config.cache.set(key, d.current)

@pytest.fixture(autouse=True)
def check_duration(request, duration_cache):
    d = duration_cache
    nodeid = request.node.nodeid
    start_time = datetime.datetime.now()
    yield
    duration = (datetime.datetime.now() - start_time).total_seconds()
    d.current[nodeid] = duration
    if d.last.get(nodeid, None) is not None:
        errorstring = "test duration over 2x last duration"
        assert duration <= (d.last[nodeid] * 2), errorstring
```

duration_cache 的作用范围是会话级别的。在所有测试用例运行之前，它会读取之前的 cache 记录（如果没有记录，就是一个空字典）。在上面的代码中，我们把读取后的字典和一个空字典都存储在名为 Duration 的 namedtuple 中，并使用 current 和 last 来访问之。之后将这个 namedtuple 传递给 check_duration，check_duration 的作用范围是函数级别的。当测试用例运行时，相同的 namedtuple 被传递给每个测试用例。当前测试的运行时间被存储在 d.current 字典里。测试结束后，汇总的 current 字典被保存在 cache 里。

多次运行后，让我们来看看 cache。

```
$ pytest -q --cache-clear test_slower_2.py
.....
5 passed in 2.80 seconds
$ pytest -q --tb=no test_slower_2.py
...E..E
5 passed, 2 error in 1.97 seconds
$ pytest -q --cache-show
------------------------ cache values ------------------------
cache/lastfailed contains:
  {'test_slower_2.py::test_slow_stuff[2]': True,
   'test_slower_2.py::test_slow_stuff[4]': True}
duration/testdurations contains:
  {'test_slower_2.py::test_slow_stuff[0]': 0.145404,
   'test_slower_2.py::test_slow_stuff[1]': 0.199585,
   'test_slower_2.py::test_slow_stuff[2]': 0.696492,
   'test_slower_2.py::test_slow_stuff[3]': 0.202118,
   'test_slower_2.py::test_slow_stuff[4]': 0.657917}

no tests ran in 0.01 seconds
```

现在看起来好多了。

4.4 使用 capsys
Using capsys

pytest 内置的 capsys 有两个功能：允许使用代码读取 stdout 和 stderr；可以临时禁止抓取日志输出。让我们先查看读取 stdout 和 stderr 的例子。

假设某个函数要把欢迎信息输出到 stdout。

ch4/cap/test_capsys.py
```python
def greeting(name):
    print('Hi, {}'.format(name))
```

你不能使用返回值来测试它，只能测试 stdout，这时可以使用 capsys 来测试。

ch4/cap/test_capsys.py
```python
def test_greeting(capsys):
    greeting('Earthling')
    out, err = capsys.readouterr()
    assert out == 'Hi, Earthling\n'
    assert err == ''

    greeting('Brian')
    greeting('Nerd')
    out, err = capsys.readouterr()
    assert out == 'Hi, Brian\nHi, Nerd\n'
    assert err == ''
```

读取到的 stdout 和 stderr 信息是从 capsys.redouterr() 中获取的。返回值从测试函数运行后捕获，或者从上次调用中获取。

上面的例子仅使用了 stdout。让我们看看另外一个使用 stderr 的例子。

ch4/cap/test_capsys.py
```python
def yikes(problem):
    print('YIKES! {}'.format(problem), file=sys.stderr)

def test_yikes(capsys):
```

```
yikes('Out of coffee!')
out, err = capsys.readouterr()
assert out == ''
assert 'Out of coffee!' in err
```

pytest 通常会抓取测试用例及被测试代码的输出。仅当全部测试会话运行结束后，抓取到的输出才会随着失败的测试显示出来。--s 参数可以关闭这个功能，在测试仍在运行期间就把输出直接发送到 stdout。通常这很方便，但是有时你可能又需要其中的部分信息。此时可以使用 capsys，capsys.disabled() 可以临时让输出绕过默认的输出捕获机制。

下面是一个例子。

ch4/cap/test_capsys.py
```
def test_capsys_disabled(capsys):
    with capsys.disabled():
        print('\nalways print this')
    print('normal print, usually captured')
```

现在每次都会显示'always print this'这条信息。

```
$ cd /path/to/code/ch4/cap
$ pytest -q test_capsys.py::test_capsys_disabled
always print this
.
1 passed in 0.01 seconds
$ pytest -q -s test_capsys.py::test_capsys_disabled
always print this
normal print, usually captured
.
1 passed in 0.00 seconds
```

正如你所看到的，不管有没有捕获输出，始终都会显示'always print this'。这是因为它是在含有 capsys.disabled() 的代码块中执行的。其他的打印语句是正常命令，所以'normal print, usally captured'这条信息仅当我们传入-s 标识进才会显示。-s 标识是-capture=no 的简写，表示关闭输出捕获。

4.5 使用 monkeypatch
Using monkeypatch

monkey patch 可以在运行期间对类或模块进行动态修改。在测试中，monkey patch 常用于替换被测试代码的部分运行环境，或者将输入依赖或输出依赖替换成更容易测试的对象或函数。pytest 内置的 `monkeypatch` 允许你在单一测试的环境里做这些事情。测试结束后，无论结果是通过还是失败，代码都会复原（所有修改都会撤销）。下面还是通过具体的例子来说明吧。通过查阅 API，我们知道了如何在测试代码中使用 monkeypatch。

monkeypatch 提供以下函数。

- `setattr(target,name,value=<notset>,raising=True)`：设置一个属性。
- `delattr(target,name=<notset>,raising=True)`：删除一个属性。
- `setitem(dic,name,value)`：设置字典中的一条记录。
- `delitem(dic,name,raising=True)`：删除字典中的一条记录。
- `setenv(name,value,prepend=None)`：设置一个环境变量。
- `delenv(name,raising=True)`：删除一个环境变量。
- `syspath_prepend(path)`：将路径 path 加入 sys.path 并放在最前，sys.path 是 Python 导入的系统路径列表。
- `chdir(path)`：改变当前的工作目录。

`raising` 参数用于指示 pytest 是否在记录不存在时抛出异常。`setenv()` 函数里的 prepend 参数可以是一个字符，如果这样设置的话，那么环境变量的值就是 value + prepend + <old value>。

为了理解 monkeypatch 的实际应用方式，我们先来看看用于生成配置文件的代码。用户目录中的配置文件存储了一些预设置信息，它们将影响程序的行为。以下代码用于读/写配置文件。

ch4/monkey/cheese.py
```python
import os
import json

def read_cheese_preferences():
    full_path = os.path.expanduser('~/.cheese.json')
    with open(full_path, 'r') as f:
    prefs = json.load(f)
    return prefs

def write_cheese_preferences(prefs):
    full_path = os.path.expanduser('~/.cheese.json')
    with open(full_path, 'w') as f:
        json.dump(prefs, f, indent=4)

def write_default_cheese_preferences():
    write_cheese_preferences(_default_prefs)
_default_prefs = {
    'slicing': ['manchego', 'sharp cheddar'],
    'spreadable': ['Saint Andre', 'camembert',
                   'bucheron', 'goat', 'humbolt fog', 'cambozola'],
    'salads': ['crumbled feta']
}
```

write_default_cheese_preferences()函数既不含参数，又没有返回值，那么该如何测试呢？它在当前用户目录中编写了一个文件，我们可以利用这一点从侧面测试。

一种办法是直接运行代码，检查文件的生成情况。在我们足够信任 read_cheese_preferences()函数测试结果的前提下，可以直接把它运用到 write_default_cheese_preferences()函数的测试里。

ch4/monkey/test_cheese.py
```python
def test_def_prefs_full():
    cheese.write_default_cheese_preferences()
    expected = cheese._default_prefs
    actual = cheese.read_cheese_preferences()
    assert expected == actual
```

这里有一个问题，用户运行此测试后，预设置文件会被覆盖，这显然不合适。

如果用户设置了 HOME 变量，那么 os.path.expanduser()函数会把~替换为

HOME 环境变量的值。让我们创建一个临时目录并将 HOME 指向它。

ch4/monkey/test_cheese.py
```
def test_def_prefs_change_home(tmpdir, monkeypatch):
    monkeypatch.setenv('HOME', tmpdir.mkdir('home'))
    cheese.write_default_cheese_preferences()
    expected = cheese._default_prefs
    actual = cheese.read_cheese_preferences()
    assert expected == actual
```

看起来不错，但其中的 HOME 变量依赖于操作系统。查阅 Python 官方文档，可以在 os.path.expanduser() 的介绍中找到这样一句话："On Windows, HOME and USERPROFILE will be used if set, otherwise a combination of...."[1]。很遗憾，这个测试不适合 Windows 用户。看来我们应该换一种实现方式。

让我们用 expanduser 替换 HOME 环境变量。

ch4/monkey/test_cheese.py
```
def test_def_prefs_change_expanduser(tmpdir, monkeypatch):
    fake_home_dir = tmpdir.mkdir('home')
    monkeypatch.setattr(cheese.os.path, 'expanduser',
                        (lambda x: x.replace('~', str(fake_home_dir))))
    cheese.write_default_cheese_preferences()
    expected = cheese._default_prefs
    actual = cheese.read_cheese_preferences()
    assert expected == actual
```

在测试中，cheese 模块中调用的 os.path.expanduser() 函数会被 lambda 表达式替换。原先该函数使用正则表达式模块的 re.sub() 函数，将~替换为我们新建的临时目录。现在已经使用了 setenv() 和 setattr() 函数来修改环境变量和属性。下面让我们使用 setitem() 函数。

有可能文件已经存在，所以要确保当 write_default_cheese_preferences() 被调用时，文件会被默认内容覆盖。

---

[1] https://docs.python.org/3.6/library/os.path.html#os.path.expanduser

ch4/monkey/test_cheese.py
```python
def test_def_prefs_change_defaults(tmpdir, monkeypatch):
    # write the file once
    fake_home_dir = tmpdir.mkdir('home')
    monkeypatch.setattr(cheese.os.path, 'expanduser',
                        (lambda x: x.replace('~', str(fake_home_dir))))
    cheese.write_default_cheese_preferences()
    defaults_before = copy.deepcopy(cheese._default_prefs)

    # change the defaults
    monkeypatch.setitem(cheese._default_prefs, 'slicing', ['provolone'])
    monkeypatch.setitem(cheese._default_prefs, 'spreadable', ['brie'])
    monkeypatch.setitem(cheese._default_prefs, 'salads',['pepper jack'])
    defaults_modified = cheese._default_prefs
    # write it again with modified defaults
    cheese.write_default_cheese_preferences()

    # read, and check
    actual = cheese.read_cheese_preferences()
    assert defaults_modified == actual
    assert defaults_modified != defaults_before
```

由于 _default_prefs 是字典，所以可以在测试运行时用 monkeypatch.setitem()来修改字典中的条目。

我们使用过 setenv()、setattr()和 setitem()。有关 del 的几个函数在形式上与 set 非常相似，只不过它们是用来删除环境变量、属性和字典条目。最后的两个 monkeypatch 函数是有关路径操作的。

syspath_prepend(path)在 sys.path 列表前加入一条路径，这可以提高你的新路径在模块搜索时的优先级。比如你可以采用这个方法，使用自定义的包、模块替换原先作用于系统范围的版本，接着使用 monkeypatch.syspatch_prepend()函数来加入含有新版本模块的路径，这样，要测试的代码就会使用新版本的模块。

chdir(path)可以在测试运行时改变当前的工作目录。这对于测试命令行脚本和其他依赖于当前目录的工具都很有用。你可以设置一个临时目录，然后使用 monkeypatch.chdir(the_tmpdir)。

还可以使用 monkeypatch 和 unittest.mock 替换模拟对象的属性值。这些内容将在第 7 章介绍。

4.6　使用 doctest_namespace
Using doctest_namespace

doctest 模块是 Python 标准库的一部分，借助它，可以在函数的文档字符串中放入示例代码，并通过测试确保有效。你可以使用 --doctest-modules 标识搜寻并运行 doctest 测试用例。在构建被标注为 autouse 的 fixture 时，可以使用内置的 doctest_namespace，这能够使 doctest 中的测试用例在运行时识别某些作用于 pytest 命名空间的字符标识，从而增强文档字符串的可读性。doctest_namespace 经常被用来在命名空间导入模块，尤其是 Python 喜欢给包或模块取别名。比如，numpy 常常是这样导入的：import numpy as np。

让我们看一个例子。模块 unnecessary_math.py 有两个函数：multiply() 和 divide()，我们希望每个人都清楚地了解这两个函数。所以在文件和函数的文档字符串中都加入了一些使用例子。

ch4/dt/1/unnecessary_math.py
```
"""
This module defines multiply(a, b) and divide(a, b).

>>> import unnecessary_math as um

Here's how you use multiply:
>>>um.multiply(4, 3)
12
>>>um.multiply('a', 3)
'aaa'
Here's how you use divide:
>>>um.divide(10, 5)
2.0
"""
```

4.6 使用 doctest_namespace

```python
def multiply(a, b):
    """
    Returns a multiplied by b.
    >>>um.multiply(4, 3)
    12
    >>>um.multiply('a', 3)
    'aaa'
    """
    return a * b

def divide(a, b):
    """
    Returns a divided by b.
    >>>um.divide(10, 5)
    2.0
    """
    return a / b
```

unnecessary_math 名字太长了，我们决定使用 um 来代替它，所以在文档顶部使用了 import unnecessary_math as um。后面的文档字符串里的代码不包含 import 语句，但一直在使用 um。问题是 pytest 将每个字符串里的代码看成是不同的测试用例，顶部的 import 语句可以保证第一个例子通过，但是后面的会失败。

```
$ cd /path/to/code/ch4/dt/1
$ pytest -v --doctest-modules --tb=short unnecessary_math.py
======================= test session starts =======================
collected 3 items

unnecessary_math.py::unnecessary_math PASSED
unnecessary_math.py::unnecessary_math.divide FAILED
unnecessary_math.py::unnecessary_math.multiply FAILED
============================ FAILURES =============================
_____ [doctest] unnecessary_math.divide _____
031
032     Returns a divided by b.
033
034     >>>um.divide(10, 5)
UNEXPECTED EXCEPTION: NameError("name 'um' is not defined",)
Traceback (most recent call last):
...
  File "<doctest unnecessary_math.divide[0]>", line 1, in <module>

NameError: name 'um' is not defined
/path/to/code/ch4/dt/1/unnecessary_math.py:34: UnexpectedException
```

```
_____ [doctest] unnecessary_math.multiply _____
022
023 >>>um.multiply(4, 3)
UNEXPECTED EXCEPTION: NameError("name 'um' is not defined",)
Traceback (most recent call last):
...
  File "<doctest unnecessary_math.multiply[0]>", line 1, in <module>
NameError: name 'um' is not defined
/path/to/code/ch4/dt/1/unnecessary_math.py:23: UnexpectedException
================= 2 failed, 1 passed in 0.03 seconds =================
```

一种解决方法是在每个文档字符串中加入 import 语句。

ch4/dt/2/unnecessary_math.py
```python
def multiply(a, b):
    """
    Returns a multiplied by b.

    >>> import unnecessary_math as um
    >>>um.multiply(4, 3)
    12
    >>>um.multiply('a', 3)
    'aaa'
    """
    return a * b

def divide(a, b):
    """
    Returns a divided by b.

    >>> import unnecessary_math as um
    >>>um.divide(10, 5)
    2.0
    """
    return a / b
```

这样做肯定能解决问题。

```
$ cd /path/to/code/ch4/dt/2
$ pytest -v --doctest-modules --tb=short unnecessary_math.py
======================= test session starts =======================
collected 3 items

unnecessary_math.py::unnecessary_math PASSED
unnecessary_math.py::unnecessary_math.divide PASSED
unnecessary_math.py::unnecessary_math.multiply PASSED
===================== 3 passed in 0.03 seconds =====================
```

但是，这样做分割了文档字符串，对读者阅读代码也没有什么实际帮助。

在顶层的 conftest.py 中使用内置的 doctest_namesapce，构建标记为 autouse 的 fixture，就可以解决之前的问题而且不用修改代码。

ch4/dt/3/conftest.py
```python
import pytest
import unnecessary_math

@pytest.fixture(autouse=True)
def add_um(doctest_namespace):
    doctest_namespace['um'] = unnecessary_math
```

pytest 会把 um 添加到 doctest_namespace 中，并把它作为 unnecessary_math 模块的别名。这样设置 conftest.py 之后，在 conftest.py 的作用范围内的任意一个 doctest 测试用例都可以使用 um。

第 7 章将详细介绍在 pytest 中运行 doctest。

4.7 使用 recwarn
Using recwarn

内置的 recwarn 可以用来检查待测代码产生的警告信息。在 Python 里，你可以添加警告信息，它们很像断言，但是并不阻止程序运行。例如，假定我们想停止支持一个原本不该发布的函数，则可以在代码里设置警告信息。

ch4/test_warnings.py
```python
import warnings
import pytest

def lame_function():
    warnings.warn("Please stop using this", DeprecationWarning)
    # rest of function
```

可以使用下面的测试用例来确保警告信息显示正确。

ch4/test_warnings.py
```python
def test_lame_function(recwarn):
    lame_function()
    assert len(recwarn) == 1
    w = recwarn.pop()
    assert w.category == DeprecationWarning
    assert str(w.message) == 'Please stop using this'
```

recwarn 的值就像是一个警告信息列表，列表里的每个警告信息都有 4 个属性 category、message、filename、lineno，从上面的代码中可以看到。

警告信息在测试开始后收集。如果你在意的警告信息出现在测试尾部，则可以在信息收集前使用 recwarn.clear() 清除不需要的内容。

除了 recwarn，pytest 还可以使用 pytest.warns() 来检查警告信息。

ch4/test_warnings.py
```python
def test_lame_function_2():
    with pytest.warns(None) as warning_list:
        lame_function()

    assert len(warning_list) == 1
    w = warning_list.pop()
    assert w.category == DeprecationWarning
    assert str(w.message) == 'Please stop using this'
```

pytest.warns() 上下文管理器可以优雅地标识哪些代码需要检查警告信息。recwarn 提供了相似的功能。读者可以自行选择。

4.8 练习
Exercises

1. 在 ch4/cache/test_slower.py 文件里有一个标记为 autouse 的 check_duration()。请把它复制到 ch3/tasks_proj/tests/conftest.py。

2. 运行第 3 章的测试用例。

3. 对于那些运行得很快的测试用例，两倍耗时仍然很快。修改 fixture 的检查时间，在两倍耗时的基础上加上 0.1 秒的延时。

4. 运行修改后的 fixture。结果是否合理？

4.9 预告
What's Next

本章介绍了很多 pytest 内置的 fixture。第 5 章将详细讲解插件。如果深入研究大型插件，完全可以再写一本书，我们主要讲解小巧好用的、自定义的 pytest 插件。

第 5 章

插件
Plugins

pytest 的自带功能已经很强大，通过添加插件可以让它变得更强大。pytest 的代码结构适合定制和扩展插件，可以借助 hook 函数来实现这一点。

如果你完成了前几章的练习，实际上就已经写过一些 pytest 插件。把 fixture 函数或 hook 函数添加到 conftest.py 文件里，就已经创建了一个本地 conftest 插件。你可以很容易把这些 conftest.py 文件转换为可安装的插件，再与其他人共享。

本章先介绍如何寻找第三方插件。现成的插件很多，也许你想写的插件已经有人写过。有很多插件是开源的，你可以进一步完善已有的插件，建立自己的分支，或者参考别人的插件来编写自己的插件。本章主要介绍如何创建自己的插件，附录 C 列出了一些已有的插件。

本章讲解如何测试、打包、发布插件。Python 的打包和发布可以写一本书，我们只挑重点介绍。我会介绍一些快捷方式，借助 PyPI 来快速发布插件。

5.1 寻找插件
Finding Plugins

你可以在好几个地方找到第三方的 pytest 插件。附录 C 中的插件都可以从 PyPI 下载，但 PyPI 不是下载 pytest 插件的唯一地方。

https://docs.pytest.org/en/latest/plugins.html

这是 pytest 网站的一个页面，用来讨论安装和使用 pytest 插件，也列出了一些常见的插件。

https://pypi.python.org

python 包索引(PyPI)上面有大量的 python 包，也有很多 pytest 插件。你可以使用 pytest、pytest-、-pytest 作为搜索关键字来搜索插件，这是因为大多数 pytest 插件的名字要么以 pytest 开始，要么以-pytest 结尾。

https://github.com/pytest-dev

GitHub 上的 pytest-dev 组保存了 pytest 源代码。这里也能找到最常用的 pytest 插件，它们是由 pytest 核心团队维护的。

5.2 安装插件
Installing Plugins

就像安装其他 Python 包一样，pytest 插件也使用 pip 安装。你可以使用几种不同的方式安装插件。

从 PyPI 安装
Install from PyPI

PyPI 是 pip 的默认仓库，从 PyPI 安装插件最简单。先让我们安装 pytest-cov 插件。

```
$ pip install pytest-cov
```

这条命令从 PyPI 安装最新的稳定版本。

从 PyPI 安装指定版本
Install a Particular Version from PyPI

如果要安装指定版本的插件，则可以用==指定版本。

```
$ pip install pytest-cov==2.4.0
```

从 .tar.gz 或 .whl 文件安装
Install from a .tar.gz or .whl File

PyPI 上的软件包一般是以.tar.gz 或.whl 为扩展名的压缩文件。这些文件通常被称为 tar 包和 wheels。如果无法使用 pip 从 PyPI 下载插件（如被防火墙屏蔽等），那么可以下载.tar.gz 或者.whl 文件，然后再安装。

这些文件不用解压就可直接使用 pip 安装。

```
$ pip install pytest-cov-2.4.0.tar.gz
```

或者

```
$ pip install pytest_cov-2.4.0-py2.py3-none-any.whl
```

从本地目录安装
Install from a Local Directory

如果插件是以.tar.gz 或.whl 的格式保存在本地或共享目录里，也可以从本地目录安装。

```
$ mkdir some_plugins
$ cp pytest_cov-2.4.0-py2.py3-none-any.whl some_plugins/
$ pip install --no-index --find-links=./some_plugins/ pytest-cov
```

`--no-index` 参数告诉 pip 不要连接到 PyPI。`--find-links=/some_plugins/`告诉 pip 查找名为 `some_plugins` 的目录。如果本地存有第三方和自定义的插件，或者你正在创建新的虚拟环境来使用持续集成或 tox（第 7 章将讨论 tox 和持续集成），这个方法尤其有用。

注意，使用本地目录安装可以通过添加==指定版本号来安装多个版本。

```
$ pip install --no-index --find-links=./some_plugins/ pytest-cov==2.4.0
```

从 Git 存储仓库安装
Install from a Git Repository

也可以直接从 Git 存储仓库安装插件。

```
$ pip install git+https: //github.com/pytest-dev/pytest-cov
```

还可以指定版本标记。

```
$ pip install git+https: //github.com/pytest-dev/pytest-cov@v2.4.0
```

或者指定一个分支。

```
$ pip install git+https: //github.com/pytest-dev/pytest-cov@master
```

从 Git 存储仓库安装特别适合使用 Git 存储自己的代码，或者 PyPI 里没有需要的插件的情况。

5.3 编写自己的插件
Writing Your Own Plugins

使用第三方的插件可以节省我们自己开发的时间，但是有些特定的测试工作需要特殊的 fixture 来完成。即使是在几个项目之间共享的 fixture，也可以通过创建一个插件来更容易地共享。你可以开发和发布自己的插件，在多个项目之间同步代码，这并不难。本节将对 pytest 做一个小小的修改，将其打包为插件，测试并学习如何发布。

插件可以包含改变 pytest 行为的 hook 函数。开发 pytest 的目的之一就是用插件改变 pytest 的运行方式。可用的 hook 函数很多，它们的详细定义可以参考 pytest 文档[1]。

我们的示例将创建一个插件来更改测试状态的显示方式。它将提供一个新的命令参数来打开这个功能，同时会增加一些对应的输出内容。具体来说，我们会将错误信息修改成 OPPORTUNITY for improvement，同时将 F 的状态修改成 O，并在标题中添加 Thanks for running the tests。下面将使用 --nice 选项来打开新增功能。

为了简便起见，我们会直接修改 conftest.py 文件。你不必以这种方式开始开发插件。常见的情况是这样的：你对某个特定项目做了修改，后来发现这些代码很有用，很值得做成插件分享。所以，我们先向 conftest.py 文件添加功能，然后将代码打包。

让我们回到 Tasks 项目。在第 2.3 节编写了一些测试，以确保调用 API 不正确时会抛出异常。但是，看起来我们错过了几个可能的错误条件。

下面是几个测试用例。

```
ch5/a/tasks_proj/tests/func/test_api_exceptions.py
import pytest
import tasks
from tasks import Task

@pytest.mark.usefixtures('tasks_db')
class TestAdd():
    """Tests related to tasks.add()."""

    def test_missing_summary(self):
        """Should raise an exception if summary missing."""
        with pytest.raises(ValueError):
            tasks.add(Task(owner='bob'))

    def test_done_not_bool(self):
        """Should raise an exception if done is not a bool."""
        with pytest.raises(ValueError):
            tasks.add(Task(summary='summary', done='True'))
```

---

[1] http://doc.pytest.org/en/latest/_modules/_pytest/hookspec.html

让我们看看它们能否通过测试。

```
$ cd /path/to/code/ch5/a/tasks_proj
$ pytest
===================== test session starts ======================
collected 57 items

tests/func/test_add.py...
tests/func/test_add_variety.py...........................
tests/func/test_add_variety2.py............
tests/func/test_api_exceptions.py.F......
tests/func/test_unique_id.py.
tests/unit/test_task.py ....
=========================== FAILURES ===========================
_____ TestAdd.test_done_not_bool _____
self = <func.test_api_exceptions.TestAdd object at 0x103a71a20>
    def test_done_not_bool(self):
        """Should raise an exception if done is not a bool."""
        withpytest.raises(ValueError):
>           tasks.add(Task(summary='summary', done='True'))
E           Failed: DID NOT RAISE <class 'ValueError'>
tests/func/test_api_exceptions.py: 20: Failed
============= 1 failed, 56 passed in 0.28 seconds ==============
```

使用详细信息选项-v再次运行。看完回溯信息后，可以把它关掉（使用选项--tb = no）。

现在,让我们使用-k TestAdd 选项来查看一个新的测试用例。可以这样做是因为没有其他测试用例名字里包含 TestAdd。

```
$ cd /path/to/code/ch5/a/tasks_proj/tests/func
$ pytest -v --tb=no test_api_exceptions.py -k TestAdd
===================== test session starts ======================
collected 9 items

test_api_exceptions.py: : TestAdd: : test_missing_summary PASSED
test_api_exceptions.py: : TestAdd: : test_done_not_bool FAILED
====================== 7 tests deselected ======================
======= 1 failed, 1 passed, 7 deselected in 0。07 seconds =======
```

我们可以尝试修复这个测试（稍后再介绍），现在先想办法让失败信息变得更"友好"。

首先，将 thank you 消息添加到标题中，这里可以使用 pytest 的 pytest_report_header()函数。

ch5/b/tasks_proj/tests/conftest.py
```python
def pytest_report_header():
    """Thank tester for running tests."""
    return "Thanks for running the tests."
```

示例显示的信息没有实际意义，但是你可以在这里添加很多有用的信息，比如用户名、指定使用的硬件、正在测试的版本信息等。所有你能转换为字符串的东西都可以放到标题头部。

接下来将更改测试的状态报告，将 F 更改为 O，将 FAILED 改为 OPPORTUNITY for improvement。hook 函数 pytest_report_teststatus()提供了这种功能。

ch5/b/tasks_proj/tests/conftest.py
```python
def pytest_report_teststatus(report):
    """Turn failures into opportunities."""
    if report.when == 'call' and report.failed:
        return (report.outcome, 'O', 'OPPORTUNITY for improvement')
```

现在得到了期望的输出结果。下面是一个不带--verbose 选项的测试输出结果，显示了一个结果为 O 的测试失败用例。

```
$ cd /path/to/code/ch5/b/tasks_proj/tests/func
$ pytest --tb=no test_api_exceptions.py -k TestAdd
===================== test session starts =====================
Thanks for running the tests.
collected 9 items

test_api_exceptions.py .O
===================== 7 tests deselected =====================
======= 1 failed, 1 passed, 7 deselected in 0.06 seconds =======
```

打开-v 或--verbose 选项可以看得更清楚。

```
$ pytest -v --tb=no test_api_exceptions.py -k TestAdd
===================== test session starts =====================
Thanks for running the tests.
collected 9 items
```

```
test_api_exceptions.py::TestAdd::test_missing_summary PASSED
test_api_exceptions.py::TestAdd::test_done_not_bool OPPORTUNITY
for improvement
====================== 7 tests deselected ======================
======= 1 failed, 1 passed, 7 deselected in 0.07 seconds =======
```

我们要做的最后一个改动是添加命令行选项--nice，只有打开--nice选项时才能使用我们自己的状态信息。

ch5/c/tasks_proj/tests/conftest.py
```python
def pytest_addoption(parser):
    """Turn nice features on with --nice option."""
    group = parser.getgroup('nice')
    group.addoption("--nice", action="store_true",
                    help="nice: turn failures into opportunities")


def pytest_report_header():
    """Thank tester for running tests."""
    if pytest.Config.getoption('nice'):
        return "Thanks for running the tests."


def pytest_report_teststatus(report):
    """Turn failures into opportunities."""
    if report.when == 'call':
        if report.failed and pytestconfig.getoption('nice'):
            return (report.outcome, 'O', 'OPPORTUNITY for improvement')
```

这里只使用了几个 hook 函数，在 pytest 的文档里可以找到更多的 hook 函数[1]。

可以手动测试这个插件，用它来测试我们的示例文件。先不使用--nice选项，以保证显示原有的内容。

```
$ cd /path/to/code/ch5/c/tasks_proj/tests/func
$ pytest --tb=no test_api_exceptions.py -k TestAdd
====================== test session starts ======================
collected 9 items

test_api_exceptions.py .F
====================== 7 tests deselected ======================
======= 1 failed, 1 passed, 7 deselected in 0.07 seconds =======
```

---

[1] https://docs.pytest.org/en/latest/writing_plugins.html

然后打开--nice 选项。

```
$ pytest --nice --tb=no test_api_exceptions.py -k TestAdd
===================== test session starts =====================
Thanks for running the tests.
collected 9 items

test_api_exceptions.py .O
===================== 7 tests deselected =====================
======= 1 failed, 1 passed, 7 deselected in 0.07 seconds =======
```

现在同时使用--nice 选项和--verbose 选项。

```
$ pytest -v --nice --tb=no test_api_exceptions.py -k TestAdd
===================== test session starts =====================
Thanks for running the tests.
collected 9 items

test_api_exceptions.py: : TestAdd: : test_missing_summary PASSED
test_api_exceptions.py: : TestAdd: : test_done_not_bool OPPORTUNITY
for improvement
===================== 7 tests deselected =====================
======= 1 failed, 1 passed, 7 deselected in 0.06 seconds =======
```

太棒了！我们只在 conftest.py 文件里添加了十几行代码，就实现了想要的功能。接下来将这些代码打包成插件。

5.4 创建可安装插件
Creating an Installable Plugin

共享插件的流程很清楚。如果你还没有在 PyPI 上共享过插件，那么不妨试一试。熟悉这个流程有助于你更轻松地阅读开源插件的代码，从而判断它们是否对你有用。

本书不介绍 Python 的打包和发布知识，读者可以在 Python 文档中查找到相应的内容[1]。不过，将刚刚创建的本地配置插件变成一个可安装的 pip 插件

---

[1] http://python-packaging.readthedocs.io 和 https://www.pypa.io

并不困难。

首先,创建一个新目录用来存放插件代码,目录可以任意取名,我们不妨将它命名为 pytest-nice。我们将在这个新目录中添加两个文件 pytest_nice.py 和 setup.py(第 5.5 节会介绍 tests 目录)。

```
pytest-nice
├── LICENCE
├── README.rst
├── pytest_nice.py
├── setup.py
└── tests
    ├── conftest.py
    └── test_nice.py
```

我们将 conftest.py 中的相关代码转移到 pytest_nice.py 文件里,同时将它们从 tasks_proj/tests/conftest.py 中删除。

`ch5/pytest-nice/pytest_nice.py`
```python
"""Code for pytest-nice plugin."""

import pytest

def pytest_addoption(parser):
    """Turn nice features on with --nice option."""
    group = parser.getgroup('nice')
    group.addoption("--nice", action="store_true",
                    help="nice: turn FAILED into OPPORTUNITY for improvement")

def pytest_report_header():
    """Thank tester for running tests."""
    if pytest.config.getoption('nice'):
        return "Thanks for running the tests."

def pytest_report_teststatus(report):
    """Turn failures into opportunities."""
    if report.when == 'call':
        if report.failed and pytest.Config.getoption('nice'):
            return (report.outcome, 'O', 'OPPORTUNITY for improvement')
```

我们需要在 setup.py 文件里调用 setup()。

ch5/pytest-nice/setup.py
```
"""Setup for pytest-nice plugin."""

from setuptools import setup

setup(
    name='pytest-nice',
    version='0.1.0',
    description='A pytest plugin to turn FAILURE into
OPPORTUNITY',
    url=
'https: //wherever/you/have/info/on/this/package',
    author='Your Name',
    author_email='your_email@somewhere.com',
    license='proprietary',
    py_modules=['pytest_nice'],
    install_requires=['pytest'],
    entry_points={'pytest11': ['nice = pytest_nice', ], },
)
```

如果你希望公开发布插件，那么最好在 setup.py 文件里提供更多有用的信息。当然，如果只是给自己和小团队用，这些就够了。

你可以在 setup() 函数中加入更多参数，这里只使用了必要的字段。version 字段记录此插件的版本，这由你自己决定。url 字段可以不填，但你会收到警告提示。author（作者）和 author_email（作者邮件）字段可以使用 maintainer（维护者）和 maintainer_email（维护者邮件）代替。license 字段域是一个简短的文本字段，用来填写开源许可方式或者公司名称等信息。py_modules 字段列出了 pytest_nice 模块，这也是这个插件里的唯一模块。这个列表可以包含不止一个模块，但是，如果有多个模块，我通常会使用 packages，而不是把所有的模块放在一个目录里。

到目前为止，setup()函数里的所有参数都是标准的，可用于所有 Python 的安装程序。pytest 插件的不同之处在于 entry_points 参数。我们使用了 entry_points= {'pytest11': ['nice = pytest_nice',],},。其中，entry_points 是 setuptools 的标准功能；pytest11 是一个特殊的标识符。通过这个设置，可以告诉 pytest 插件的名称是 nice，模块的名称是

pytest_nice。如果使用了包，则设置可以改成：

entry_points={'pytest11 ' : [' name_of_plugin = myprojectpluginmodule ',],},

再来看看 README.rst（自述文件）。setuptools 要求必须提供 README 文件。如果没有提供，则会收到警告。

```
...
warning: sdist: standard file not found: should have one of README,
   README.rst, README.txt
...
```

使用 README 文件展示项目相关信息是一个好习惯。下面是我为 pytest-nice 写的 README 文件。

ch5/pytest-nice/README.rst
```
pytest-nice: A pytest plugin
==============================
Makes pytest output just a bit nicer during failures.

Features
--------
- Includes user name of person running tests in pytest output.
- Adds ``--nice`` option that:
  - turns ``F`` to ``O``
  - with ``-v``, turns ``FAILURE`` to ``OPPORTUNITY for improvement``
Installation
------------
Given that our pytest plugins are being saved in .tar.gz form in the
shared directory PATH, then install like this:
: :
    $ pip install PATH/pytest-nice-0.1.0.tar.gz
    $ pip install --no-index --find-links PATH pytest-nice
Usage
-----
: :
    $ pytest –nice
```

README 文件可以写很多内容，我写的只是一个简单的示例。

5.5 测试插件
Testing Plugins

插件像其他代码一样，也需要测试。然而，测试工具本身的测试有点棘手。在第 5.3 节，测试插件的方式是用它手工测试示例文件，然后查看输出是否正确。也可以使用一个名为 `pytester` 的插件来自动完成同样的工作。这个插件是 pytest 自带的，但默认情况下是关闭的。

pytest-nice 的测试目录下有两个文件：conftest.py 和 test_nice.py。要使用 pytester，只需要在 conftest.py 文件里添加一行代码。

ch5/pytest-nice/tests/conftest.py
```
"""pytester is needed for testing plugins."""
pytest_plugins = 'pytester'
```

这样就开启了 pytester 插件。我们要使用的 fixture 叫 testdir，开启 pytester 后就可以使用了。

插件的测试步骤与手工测试的步骤是一样的。

1. 生成一个示例测试文件。

2. 在包含示例文件的目录中运行 pytest。

3. 检查输出。

4. 检查结果为 0，表示全部测试通过；检查结果标识为 1，表示有些失败。

让我们看一个例子：

ch5/pytest-nice/tests/test_nice.py
```
def test_pass_fail(testdir):
# create a temporary pytest test module
    testdir.makepyfile("""
        def test_pass():
            assert 1 == 1

        def test_fail():
            assert 1 == 2
```

```
    """)

    # run pytest
    result = testdir.runpytest()

    # fnmatch_lines does an assertion internally
    result.stdout.fnmatch_lines([
        '*.F', # .for Pass, F for Fail
    ])

    # make sure that that we get a '1' exit code for the testsuite
    assert result.ret == 1
```

testdir 自动创建了一个临时目录用来存放测试文件。它有一个名为 makepyfile() 的方法，允许我们写入测试文件。在上面的例子中，我们创建了两个测试用例：一个通过测试，一个失败。

现在使用 testdir.runpytest() 来运行 pytest 并测试新的测试文件。这里你还可以设置选项。返回值的类型是 RunResult[1]，便于进一步查看。

通常，我会查看标准输出 stdout 和标准返回 ret。如果是像之前我们手动检查输出结果的情况，则可以使用 fnmatch_lines 传入我们希望在输出中看到的字符串列表，同时确保测试全部通过时 ret 返回值为 0，否则返回 1。fnmatch_lines 中的字符串可以包括通配符，也可以使用示例文件进行更多测试。让我们来创建一个新的 fixture。

ch5/pytest-nice/tests/test_nice.py
```
@pytest.fixture()
def sample_test(testdir):
    testdir.makepyfile("""
        def test_pass():
            assert 1 == 1
        def test_fail():
            assert 1 == 2
    """)
    return testdir
```

剩余的测试可以使用 sample_test 作为测试文件的目录。下面是用来测试

---

[1] https://docs.pytest.org/en/latest/writing_plugins.html#_pytest.pytester.RunResult

其他选项的测试用例。

ch5/pytest-nice/tests/test_nice.py
```python
def test_with_nice(sample_test):
    result = sample_test.runpytest('--nice')
    result.stdout.fnmatch_lines(['*.O', ]) # .for
    Pass,O for Fail
    assert result.ret == 1

    def test_with_nice_verbose(sample_test):
    result = sample_test.runpytest('-v', '--nice')
    result.stdout.fnmatch_lines([
        '*: : test_fail OPPORTUNITY for improvement',
    ])
    assert result.ret == 1

def test_not_nice_verbose(sample_test):
    result = sample_test.runpytest('-v')
    result.stdout.fnmatch_lines(['*: : test_fail FAILED'])
    assert result.ret == 1
```

还要写几个测试用例。让我们先确保感谢信息会出现在页头。

ch5/pytest-nice/tests/test_nice.py
```python
def test_header(sample_test):
    result = sample_test.runpytest('--nice')
    result.stdout.fnmatch_lines(['Thanks for running the tests.'])

def test_header_not_nice(sample_test):
    result = sample_test.runpytest()
    thanks_message = 'Thanks for running the tests.'
    assert thanks_message not in result.stdout.str()
```

这个用例也可以供其他测试使用，但我希望把它单独列为一个用例，以确保一个用例只做一件事。最后查看帮助信息。

ch5/pytest-nice/tests/test_nice.py
```python
def test_help_message(testdir):
    result = testdir.runpytest('--help')

    # fnmatch_lines does an assertion internally
    result.stdout.fnmatch_lines([
        'nice:',
        '*--nice*nice: turn FAILED into OPPORTUNITY for improvement',
    ])
```

这可以进一步确保我们的插件工作正常。

运行这些测试用例之前，我们要确保插件已经安装在 pytest-nice 目录里。可以安装打包压缩的 .zip.gz 文件，或者直接在当前目录安装。

```
$ cd /path/to/code/ch5/pytest-nice/
$ pip install .
Processing /path/to/code/ch5/pytest-nice
Requirement already satisfied: pytest in
    /path/to/venv/lib/python3.6/site-packages (from
pytest-nice==0.1.0)
Requirement already satisfied: py>=1.4.33 in
    /path/to/venv/lib/python3.6/site-packages (from
pytest->pytest-nice==0.1.0)
Requirement already satisfied: setuptools in
    /path/to/venv/lib/python3.6/site-packages (from
pytest->pytest-nice==0.1.0)
Building wheels for collected packages: pytest-nice
    Running setup.py bdist_wheel for pytest-nice ... done
    ...
Successfully built pytest-nice
Installing collected packages: pytest-nice
Successfully installed pytest-nice-0.1.0
```

安装后运行测试用例。

```
$ pytest -v
===================== test session starts =====================
plugins: nice-0.1.0
collected 7 items

tests/test_nice.py: : test_pass_fail PASSED
tests/test_nice.py: : test_with_nice PASSED
tests/test_nice.py: : test_with_nice_verbose PASSED
tests/test_nice.py: : test_not_nice_verbose PASSED
tests/test_nice.py: : test_header PASSED
tests/test_nice.py: : test_header_not_nice PASSED
tests/test_nice.py: : test_help_message PASSED
=================== 7 passed in 0.34 seconds ==================
```

很好！所有测试都通过了。现在可以像卸载其他 Python 包和插件一样卸载自己的插件。

```
$ pip uninstall pytest-nice
Uninstalling pytest-nice-0.1.0:
  /path/to/venv/lib/python3.6/site-packages/pytest-nice.egg-link
  ...
Proceed (y/n)? y
  Successfully uninstalled pytest-nice-0.1.0
```

别忘了，你还可以查看 PyPI 上的其他 pytest 插件来学习测试插件的方法。

5.6 创建发布包
Creating a Distribution

插件快完工了，你可以在命令行使用 setup.py 文件创建发布文件。

```
$ cd /path/to/code/ch5/pytest-nice
$ python setup.py sdist
running sdist
running egg_info
creating pytest_nice.egg-info
...
running check
creating pytest-nice-0.1.0
...
creating dist
Creating tar archive
...
$ ls dist
pytest-nice-0.1.0.tar.gz
```

（请注意，sdist 代表源码分发。）

在 pytest-nice 的 dist 目录下现在包含一个新文件 pytest-nice-0.1.0.tar.gz。该文件可用来在任何地方安装我们的插件，甚至包括当前目录。

```
$ pip install dist/pytest-nice-0.1.0.tar.gz
Processing ./dist/pytest-nice-0.1.0.tar.gz
...
Installing collected packages: pytest-nice
Successfully installed pytest-nice-0.1.0
```

当然，你可以将 tar.gz 文件放在任何你想要的地方使用和共享。

通过共享目录分发插件
Distributing Plugins Through a Shared Directory

pip 支持从共享目录安装软件包，要通过共享目录分发我们的插件，只需要选择一个目录，然后把插件的 .tar.gz 文件放入其中。这里假设要把 pytest-nice- 0.1.0.tar.gz 放到一个叫 myplugins 的目录里。

从 myplugins 安装 pytest-nice。

`$ pip install --no-index --find-links myplugins`

--no-index 选项告诉 pip 不要查找 PyPI。

--find-links myplugins 告诉 PyPI 在 myplugins 中查找要安装的软件包。最后，pytest-nice 是我们想要安装的插件名。

如果你修复了插件的 bug，希望更新版本（myplugins 中有较新的文件），可以使用 --upgrade 选项更新。

`$ pip install --upgrade --no-index --find-links myplugins pytest-nice`

这跟通常的 pip 用法一样，只是多了两个选项：--no-index 和 --find-links myplugins。

通过 PyPI 发布插件
Distributing Plugins Through PyPI

如果你希望向全世界分享你的插件，还要做一些工作。本书不打算做进一步介绍了，建议查看 Python 用户指南中的相关介绍[1]。

另一个分享 pytest 插件的好地方是 cookiecutter 项目[2]。

```
$ pip install cookiecutter
$ cookiecutter https:
//github.com/pytest-dev/cookiecutter-pytest-plugin
```

---

[1] https://packaging.python.org/distributing
[2] https://github.com/pytest-dev/cookiecutter-pytest-plugin

cookiecutter 首先会询问插件的基本信息，然后创建一个目录以方便存放代码。该项目是由 pytest 的核心开发者维护的，他们一直在坚持更新。

5.7 练习
Exercises

在 ch4/cache/test_slower.py 文件中，有一个设置为 autouse 的 fixture 叫 check_duration()。我们在第 4 章的练习中也用到了它。现在，让我们将它做成一个插件。

1. 创建一个名为 pytest-slower 的目录，用于保存新插件的代码。

2. 将相关文件放入目录，把 pytest-slower 制作成可安装的插件。

3. 为插件编写测试代码。

4. 查看 Python 包列表[1]并搜索 pytest-，找一个你感兴趣的 pytest 插件。

5. 安装你找到的插件，并在 Tasks 项目测试中试用。

5.8 预告
What's Next

你已经与 conftest.py 文件打了很多交道，知道配置文件（如 pytest.ini）可以改变 pytest 的运行方式。第 6 章将运行不同的配置文件，学习如何让测试变得更轻松。

---

[1] https://pypi.python.org/pypi

第 6 章

配置
Configuration

到目前为止，除了第 5 章对 conftest.py 做过较深入的介绍，本书对非测试文件都只是顺带提一提。接下来将深入讲解 pytest 的配置文件，分析它们是如何改变 pytest 的运行方式的。我会继续用 Tasks 项目举例。

6.1 理解 pytest 的配置文件
Understanding pytest Configuration Files

在讨论改变 pytest 的默认运行方式之前，让我们看看 pytest 里都有哪些非测试文件。

读者也许能想到下面这几个。

- pytest.ini：pytest 的主配置文件，可以改变 pytest 的默认行为，其中有很多可配置的选项，本章大部分内容将介绍 pytest.ini 的配置。
- conftest.py：是本地的插件库，其中的 hook 函数和 fixture 将作用于该文件所在的目录以及所有子目录。第 5 章曾详细介绍过 conftest.py 文件。
- __init__.py：每个测试子目录都包含该文件时，那么在多个测试目录中可以出现同名测试文件。将在第 6.9 节看到一个因缺少 __init__.py 文件而出错的例子。

如果你使用 tox 工具，也许会用到 tox.ini。

- tox.ini：它与 pytest.ini 类似，只不过是 tox 的配置文件。你可以把 pytest 的配置都写在 tox.ini 里，这样就不用同时使用 tox.ini 和 pytest.ini 两个文件。tox 将在第 7 章详细介绍。

如果想要发布一个 Python 包(比如 Tasks)，下面这个文件很有用。

- setup.cfg：它也采用 ini 文件的格式，而且可以影响 setup.py 的行为。你可以在 setup.py 里添加几行代码，使用 python setup.py test 运行所有的 pytest 测试用例。如果你打算发布包，也可以使用 setup.cfg 文件存储 pytest 的配置信息（附录 D 将详细介绍方法）。

无论你选择哪个文件存储 pytest 的配置信息，它们的格式几乎是一样的。

pytest.ini：

ch6/format/pytest.ini
```
[pytest]
addopts = -rsxX -l --tb=short --strict
xfail_strict = true
... more options ...
```

tox.ini：

ch6/format/tox.ini
```
... tox specific stuff ...
[pytest]
addopts = -rsxX -l --tb=short --strict
xfail_strict = true
... more options ...
```

setup.cfg：

ch6/format/setup.cfg
```
... packaging specific stuff ...
[tool:pytest]
addopts = -rsxX -l --tb=short --strict
xfail_strict = true
... more options ...
```

唯一不同的是，setup.cfg 文件里的头信息是[tool:pytest]而不是[pytest]。

用 pytest --help 查看 ini 文件选项
List the Valid ini-file Options with pytest --help

可以使用 pytest-help 命令查看 pytest.ini 的所有设置选项。

```
$ pytest --help
...
[pytest] ini-options in the first pytest.ini|tox.ini|setup.cfg file found:

  markers (linelist)         markers for test functions
  norecursedirs (args)       directory patterns to avoid for recursion
  testpaths (args)           directories to search for tests when no files or
                             directories are given in the command line.
usefixtures (args)           list of default fixtures to be used with this
project
python_files (args)          glob-style file patterns for Python test
module discovery
python_classes (args)        prefixes or glob names for Python test class
discovery
  python_functions (args)    prefixes or glob names for Python test function
                             and method discovery
  xfail_strict (bool)        default for the strict parameter of xfail markers
                             when not given explicitly (default: False)
  doctest_optionflags (args) option flags for doctests
  addopts (args)             extra command line options
  minversion (string)        minimally required pytest version
...
```

本章将逐个讲解这些设置选项，只有 doctest_optionflags 例外，它留到第 7 章再讲解。

插件可以添加 ini 文件选项
Plugins Can Add ini-file Options

除了前面列出的这些选项，利用插件和 confteset.py 文件还可以添加新的选项。而且新增的选项也可以使用 pytest --help 查看。

接下来让我们看看 pytest 的 ini 文件里有哪些可以修改的设置。

6.2 更改默认命令行选项
Changing the Default Command-Line Options

我们已经用过很多 pytest 命令行选项了,比如-v/--verbose 可以输出详细信息,-l/--showlocals 可以查看失败测试用例里堆栈中的局部变量。你也许经常要用到某些选项,又不想重复输入,这时可以使用 pytest.ini 文件里的 addopts 设置。下面是我自己常用的设置。

```
[pytest]
addopts = -rsxX -l --tb=short –strict
```

--rsxX 表示 pytest 报告所有测试用例被跳过、预计失败、预计失败但实际通过的原因。-l 表示 pytest 报告所有失败测试的堆栈中的局部变量。--tb=short 表示简化堆栈回溯信息,只保留文件和行数。--strict 选项表示禁止使用未在配置文件中注册的标记。接下来会介绍标记的注册。

6.3 注册标记来防范拼写错误
Registering Markers to Avoid Marker Typos

第 2.4 节曾介绍过,自定义标记可以简化测试工作,让我们用指定的标记运行某个测试子集。但是,标记很容易拼错,比如把@pytest.mark.smoke 拼成 @pytest.mark.somke。默认情况下,这不会引起程序错误。pytest 会以为这是你创建的另一个标记。为了避免拼写错误,可以在 pytest.ini 文件里注册标记。

```
[pytest]
markers =
  smoke:  Run the smoke test functions for tasks project
  get:  Run the test functions that test tasks.get()
```

标记注册好后,可以通过 pytest --markers 来查看。

```
$ cd /path/to/code/ch6/b/tasks_proj/tests
$ pytest --markers
```

```
@pytest.mark.smoke: Run the smoke test test functions

@pytest.mark.get: Run the test functions that test tasks.get()

@pytest.mark.skip(reason=None): skip the ...
...
```

没有注册的标记不会出现在 --markers 列表里。如果使用了 --strict 选项，遇到拼写错误的标记或未注册的标记就会报错。ch6/a/tasks_proj 和 ch6/b/tasks_proj 的不同之处在于 pytest.ini 文件的内容不同。ch6/a 里的 pytest.ini 是空文件。让我们在不注册标记的情况下运行这些测试用例。

```
$ cd /path/to/code/ch6/a/tasks_proj/tests
$ pytest --strict --tb=line
===================== test session starts =====================
collected 45 items / 2 errors
========================= ERRORS ==========================
_____ ERROR collecting func/test_add.py _____
func/test_add.py:20: in <module>
    @pytest.mark.smoke
...
E AttributeError: 'smoke' not a registered marker
_____ ERROR collecting func/test_api_exceptions.py _____
func/test_api_exceptions.py:30: in <module>
    @pytest.mark.smoke
...
E AttributeError: 'smoke' not a registered marker
!!!!!!!!!!! Interrupted: 2 errors during collection !!!!!!!!!!!!
=================== 2 error in 0.24 seconds ====================
```

如果你在 pytest.ini 文件里注册了标记，那么可以同时在 addopts 里加入 --strict。这个技巧很好用，你一定会喜欢的。让我们在 Tasks 项目里加入下面的 pytest.ini 文件。

```
ch6/b/tasks_proj/tests/pytest.ini
[pytest]
addopts = -rsxX -l --tb=short --strict
markers =
  smoke: Run the smoke test test functions
  get: Run the test functions that test tasks.get()
```

这里使用了我常用的一些配置：-rsxX 用来报告哪些测试用例被跳过、预计失败、预计失败但实际通过，--tb=short 用来简化失败信息，--strict 只允许使用注册过的标记。之后就是指定项目中允许使用的标记列表。

下面来指定运行冒烟测试。

```
$ pytest --strict -m smoke
===================== test session starts =====================
collected 57 items

func/test_add.py .
func/test_api_exceptions.py ..
===================== 54 tests deselected =====================
=========== 3 passed, 54 deselected in 0.06 seconds ===========
```

6.4 指定 pytest 的最低版本号
Requiring a Minimum pytest Version

minversion 选项可以指定运行测试用例的 pytest 的最低版本。例如，测试两个浮点数的值是否非常接近，我习惯使用 approx()函数，但这个功能直到 pytest 3.0 才出现。为了避免混淆，我在使用 approx()函数的项目中增加了一行配置。

```
[pytest]
minversion = 3.0
```

如果有人使用老版本的 pytest 运行该测试，就会得到一个错误信息。

6.5 指定 pytest 忽略某些目录
Stopping pytest from Looking in the Wrong Places

pytest 执行测试搜索时，会递归遍历所有子目录，包括某些你明知道没必要遍历的目录。遇到这种情况，你可以使用 norecurse 选项简化 pytest 的搜索工作。

norecurse 的默认设置是.* build dist CVS _darcs {arch}和*.egg。因为有.*，所以将虚拟环境命名为.venv 是一个好主意，所有以.开头的目录都不会被访问。但是，我习惯将它命名为 venv，那么需要把它加入 norecursedirs 里。

就 Tasks 项目而言，可以让 pytest 忽略 src 目录，因为让 pytest 在那里搜索测试用例完全是浪费时间。

```
[pytest]
norecursedirs = .* venv src *.egg dist build
```

更改一个已经设置好的值时，可以先找到原来的默认值，并把它放在新添加的内容后面，就像我在上面写的*.egg dist build 一样。

norecursedirs 与 testpaths 密切相关，下面来看看 testpaths。

6.6 指定测试目录
Specifying Test Directory Locations

norecursedirs 告诉 pytest 哪些路径不用访问，而 testpaths 则指示 pyteset 去哪里访问。testpaths 是一系列相对于根目录的路径，用于限定测试用例的搜索范围。只有在 pytest 未指定文件目录参数或测试用例标识符时，该选项才会启用。

对于 Tasks 项目来说，假设我们把 pytest.ini 文件放在了 tasks_proj 目录，而不是之前的 tests 目录。

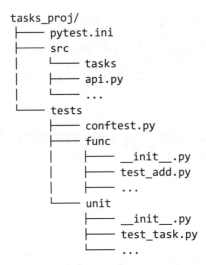

那么可以把 tests 放到 testpaths 里。

```
[pytest]
testpaths = tests
```

现在只要从 tasks_proj 开始运行 pytest，pytest 就只会访问 tasks_proj/tests。我自己在开发和调试阶段经常更换测试目录，所以没有必要配置整个搜索路径，只需要指定一个特定的文件或目录即可。因此，这个设置对我来说并没有太大帮助。

但是，这个设置对于集成测试和 tox 很有用，因为在这种情况下根目录是确定的，你可以列出相对于根目录的路径。如果你希望缩短测试时间，也可以使用这个设置，定义搜索范围可以节约测试搜索的时间。

乍一看，同时使用 testpaths 和 norecursedirs 似乎没什么意义。然而，正如你所看到的，testpaths 对于要从文件系统的不同位置进行的交互式测试没有太大帮助。这种情况下，norecursedirs 会更有用。此外，如果有一个命名为 tests 的目录却不包含任何测试文件，那么你可以使用 norecursedirs 避开这个目录。但是，在一个没有测试文件却命名为 tests 的目录下添加工作目录有什么意义呢？

6.7 更改测试搜索的规则
Changing Test Discovery Rules

pytest 根据一定的规则搜索并运行测试。标准的测试搜索规则如下。

- 从一个或多个目录开始查找。你可以在命令行指定文件名或目录名。如果未指定，则使用当前目录。
- 在该目录和所有子目录下递归查找测试模块。
- 测试模块是指文件名为 test_*.py 或*_test.py 的文件。
- 在测试模块中查找以 test_ 开头的函数名。
- 查找名字以 Test 开头的类。其中，首先筛选掉包含 __init__ 函数的类，再查找类中以 Test_ 开头的类方法。

以上是标准的测试搜索规则，你也可以更改它们。

python_classes

通常，pytest 的测试搜索规则是寻找以 Test*开头的测试类，而且这个类不能有 __init__() 函数。但是，如果把测试类命名为 <something>Test 或 <something>Suite 怎么办？python_classes 就可以解决这个问题。

```
[pytest]
python_classes = *Test Test* *Suite
```

这个设置允许我们像下面这样给类取名。

```
class DeleteSuite():
  def test_delete_1():
     ...
  def test_delete_2():
     ...
  ...
```

python_files

像 pytest_classes 一样，python_files 可以更改默认的测试搜索规则，而不是仅查找以 test_*开头的文件和以*_test 结尾的文件。

假设你有一个自定义的测试框架，其中的测试文件统一命名为 check_<something>.py。现在，你不必重命名所有的测试文件，只要在 pytest.ini 文件里增加一行配置即可。

```
[pytest]
python_files = test_* *_test check_*
```

很简单吧。现在可以保留你的命名习惯，或者选择慢慢习惯 pytest 的命名规范。

python_functions

python_functions 与之前的两个设置类似，它只是用来测试函数和方法的命名。默认命名规则以 test_*开头。如果你想添加 check_*，则只需要增加一行配置。

```
[pytest]
python_functions = test_* check_*
```

pytest 的命名规则并不是强制性的。如果你不喜欢默认的命名规则，完全可以对它进行修改。当然，你最好有一个充分的理由，比如要迁移上百个现成的测试用例。

6.8 禁用 XPASS
Disallowing XPASS

设置 xfail_strict = true 将会使那些被标记为@pytest.mark.xfail 但实际通过的测试用例也被报告为失败。我认为这个设置应该一直保持。关于 xfail 标记的用法，请参考第 2.6 节。

6.9 避免文件名冲突
Avoiding Filename Collisions

过去我一直想知道为什么每个项目的所有测试子目录下都有一个 __init__.py 文件。后来我明白了，只有这样，才能在多个目录中使用同名的测试文件。

这里有一个例子，目录 a 和目录 b 都有一个 test_foo.py 文件。这两个文件的内容如下：

```
ch6/dups/a/test_foo.py
def test_a():
    pass
```

```
ch6/dups/b/test_foo.py
def test_b():
    pass
```

目录结构如下：

```
dups
├──a
│   └──test_foo.py
└──b
    └──test_foo.py
```

虽然这两个文件的内容不同，但是它们还会产生冲突。虽然你可以单独运行它们的每一个，但是让 pytest 从 dups 目录执行就不可行了。

```
$ cd /path/to/code/ch6/dups
$ pytest a
================= test session starts =================
collected 1 items

a/test_foo.py .
=============== 1 passed in 0.01 seconds ===============
$ pytest b
================= test session starts =================
collected 1 items

b/test_foo.py .
=============== 1 passed in 0.01 seconds ===============
$ pytest
```

```
=================== test session starts ===================
collected 1 items / 1 errors
========================= ERRORS =========================
_____ ERROR collecting b/test_foo.py _____
import file mismatch:
imported module 'test_foo' has this __file__ attribute:
  /path/to/code/ch6/dups/a/test_foo.py
which is not the same as the test file we want to collect:
  /path/to/code/ch6/dups/b/test_foo.py
HINT: remove __pycache__ / .pyc files and/or use a unique basename
  for your test file modules
!!!!!!!! Interrupted: 1 errors during collection !!!!!!!!
================ 1 error in 0.15 seconds ================
```

错误信息并没有说清楚哪里出错了。

要解决这个问题，只需要在两个子目录里各添加一个空的 __init__.py 文件即可。修改后的 dups_fixed 目录结构与 dups 的目录结构一样，只是增加了两个 __init__.py 文件。

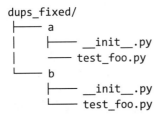

让我们再试着从 dups_fixed 目录执行测试。

```
$ cd /path/to/code/ch6/dups_fixed
$ pytest
=================== test session starts ===================
collected 2 items

a/test_foo.py .
b/test_foo.py .
=============== 2 passed in 0.01 seconds ===============
```

现在一切正常。你可能认为自己不会使用重名的文件，所以这不是一个问题。如果真的没有重名的文件，那当然很好，可是随着项目规模的增长，测试目录也会增加。你真的想在出问题的时候再来修复么？我建议从一开始就在目录里放上这个文件，把它当成一个习惯，这样你就不必再担心重名的问题了。

6.10 练习
Exercises

我们在第 5 章创建了一个 pytest-nice 插件并且包含一个 --nice 的命令行选项。现在再增加一个 pytest.ini 配置选项，命名为 nice。

1. 向 pytest_nice.py 的 hook 函数 pytest_addoption 里添加如下代码：
   Parser.addini('nice', type='bool', help='Turn failures into opportunities.')

2. 在插件里调用 getoption() 的地方也调用 getini('nice')。

3. 把 nice 添加到 pytest.ini 文件里并进行手工测试。

4. 别忘了插件测试。添加一个测试用例来确认 pytest.ini 里的 nice 设置工作正常。

5. 把这些测试用例添加到插件测试目录。你需要查阅一些额外的 pytester 功能[1]。

6.11 预告
What's Next

pytest 除了提供强大的功能（尤其在使用插件以后）外，还可以很好地与其他开发工具、测试工具配合使用。第 7 章将介绍 pytest 与其他工具的搭配用法。

---

[1] https://docs.pytest.org/en/latest/_modules/_pytest/pytester.html#Testdir

第 7 章

pytest 与其他工具的搭配使用
Using pytest with Other Tools

我们通常不会单独使用 pytest，而是与测试环境里的其他工具一起使用。本章将介绍一些经常与 pytest 搭配使用、提高测试效率的工具。我不可能介绍完所有的工具，只希望抛砖引玉，能给读者一定的启发。

7.1 pdb：调试失败的测试用例
pdb: Debugging Test Failures

pdb 是 Python 标准库里的调试模块。在 pytest 里，你可以使用 `--pdb` 选项在测试失败时开启调试。让我们看看在 Tasks 项目里如何使用 pdb。

第 3.10 节的 Tasks 项目留有几个错误。

```
$ cd /path/to/code/ch3/c/tasks_proj
$ pytest --tb=no -q
.........................................FF.FFFF
FFFFFFFFFFFFFFFFFFFFFFFFFFFFFFFFFF.FFF..........
42 failed, 54 passed in 4.74 seconds
```

在介绍 pdb 如何帮助我们调试这些错误前，让我们看看 pytest 里可以提高调试效率的选项（参考第 1.4 节）。

- --tb=[auto/long/short/line/native/no]：指定堆栈回溯信息粒度。
- -v/--verbose：显示所有的测试用例名字，测试通过情况。
- -l/--showlocals：显示错误堆栈里的局部变量。
- -lf/--last-failed：只运行上次失败的测试。
- -x/--exitfirst：当测试遇错时即停止。
- --pdb：在错误出现时开启交互式调试。

安装 MongoDB

第 3 章曾提到，运行 MongoDB 相关的测试用例需要安装 MongoDB 和 pymongo 包。我一直使用社区版的 MongoDB[1]。pymongo 包可以使用以下命令安装：`pip install pymongo`。这是本书最后一个使用 MongoDB 的测试用例。如果需要在没有 MongoDB 的情况下调试，可以切换到 code/ch2/ 目录下运行 pytest，这个目录下也有几个失败的测试用例。

我们刚刚运行了 code/ch3/c 下面的测试用例，发现部分用例失败了，但没有发现堆栈回溯以及具体测试用例的名字，这是因为我们使用了--tb=no（该参数关闭了堆栈回溯），而且--verbose 选项也没有打开。让我们打开--verbose 选项后再次运行（最多允许失败三个）。

```
$ pytest --tb=no --verbose --lf --maxfail=3
====================== test session starts ======================
run-last-failure: rerun last 42 failures
collected 96 items

tests/func/test_add.py::test_add_returns_valid_id[mongo] FAILED
tests/func/test_add.py::test_added_task_has_id_set[mongo] FAILED
tests/func/test_add_variety.py::test_add_1[mongo] FAILED
!!!!!!!!!!! Interrupted: stopping after 3 failures !!!!!!!!!!!!
====================== 54 tests deselected ======================
=========== 3 failed, 54 deselected in 3.14 seconds ============
```

---

[1] https://www.mongodb.com/download-center

现在我们知道哪些测试失败了。让我们再运行一次，这次打开-x 选项（只查看一个），不使用--tb=no（打开堆栈回溯），同时打开-l 选项（显示局部变量）。

```
$ pytest -v --lf -l -x
===================== test session starts =====================
run-last-failure: rerun last 42 failures
collected 96 items

tests/func/test_add.py::test_add_returns_valid_id[mongo] FAILED
========================== FAILURES ==========================
_____ test_add_returns_valid_id[mongo] _____
tasks_db = None

    def test_add_returns_valid_id(tasks_db):
        """tasks.add(<valid task>) should return an integer."""
        # GIVEN an initialized tasks db
        # WHEN a new task is added
        # THEN returned task_id is of type int
        new_task = Task('do something')
        task_id = tasks.add(new_task)
>       assert isinstance(task_id, int)
E       AssertionError: assert False
E        +  where False = isinstance(ObjectId('59783baf8204177f24cb1b68'), int)
new_task   = Task(summary='do something', owner=None, done=False, id=None)
task_id    = ObjectId('59783baf8204177f24cb1b68')
tasks_db = None
tests/func/test_add.py:16: AssertionError
!!!!!!!!!!!! Interrupted: stopping after 1 failures !!!!!!!!!!!!
===================== 54 tests deselected =====================
=========== 1 failed, 54 deselected in 2.47 seconds ===========
```

大多数情况下，这些信息足够我们了解测试失败的原因了。在这个例子里，很明显，task_id 不是整数，而是一个 ObjectId 类的实例。ObjectId 在 MongoDB 里用做对象标识码。我创建 taskdb_pymongo.py 这一层的目的是想隐藏 MongoDB 的实现细节。很明显，在这个例子里行不通。

下面介绍如何在 pytest 中使用 pdb。假定这个错误不是那么明显，我们让 pytest 在错误的地方启动 pdb 调试（使用--pdb 参数）。

```
$ pytest -v --lf -x --pdb
===================== test session starts ======================
run-last-failure: rerun last 42 failures
collected 96 items

tests/func/test_add.py: : test_add_returns_valid_id[mongo] FAILED
>>>>>>>>>>>>>>>>>>>>>>>>> traceback >>>>>>>>>>>>>>>>>>>>>>>>>

tasks_db = None

    def test_add_returns_valid_id(tasks_db):
        """tasks.add(<valid task>) should return an integer."""
        # GIVEN an initialized tasks db
        # WHEN a new task is added
        # THEN returned task_id is of type int
        new_task = Task('do something')
        task_id = tasks.add(new_task)
>       assert isinstance(task_id, int)
E       AssertionError: assert False
E        + where False = 
isinstance(ObjectId('59783bf48204177f2a786893'), int)

tests/func/test_add.py:16: AssertionError
>>>>>>>>>>>>>>>>>>>>>>>>> entering PDB >>>>>>>>>>>>>>>>>>>>>>>
> /path/to/code/ch3/c/tasks_proj/tests/func/test_add.py(16)
>   test_add_returns_valid_id()
-> assert isinstance(task_id, int)
(Pdb)
```

(Pdb)提示符出现了,表明可以使用 pdb 的交互调试功能。查看错误时,有以下常用命令。

- p/print expr:输出 expr 的值。

- pp expr:美化输出 expr 的值。

- l/list:列出错误并显示错误之前和之后的 5 行代码。

- l/list begin,end:列出错误,并通过行号指定需要显示的代码区域。

- a/args:打印当前函数的参数列表(断点发生在帮助函数中会很实用)。

- u/up:移动到堆栈的上一层。

- d/down:移动到堆栈的下一层。

- q/quit:退出当前调试会话。

因为我们是从 assert 语句开始的，其他控制断点移动的命令（如 step 和 next）暂时用不上。可以直接在 pdb shell 中输入变量名，从而得到变量的值。

使用 p/print expr 查看函数中的局部变量，与使用 -l/--showlocals 命令行选项的效果是类似的。

```
(Pdb) p new_task
Task(summary='do something', owner=None, done=False, id=None)
(Pdb) p task_id
ObjectId('59783bf48204177f2a786893')
(Pdb)
```

现在可以退出调试，让 pytest 继续测试。

```
(Pdb) q

!!!!!!!!!!!!! Interrupted:  stopping after 1 failures !!!!!!!!!!!!
====================== 54 tests deselected ======================
=========== 1 failed, 54 deselected in 123.40 seconds ===========
```

如果未使用 -x 选项，pytest 会在下一次失败发生时再次打开 pdb。请查阅 python 文档了解更多使用 pdb 模块的方法[1]。

7.2　coverage.py：判断测试覆盖了多少代码
coverage.py: Determining How Much Code Is Tested

测试覆盖率是指项目代码被测试用例覆盖的百分比。我们在测试 Tasks 项目时，有些 Tasks 功能被测试用例覆盖，有些则没有。覆盖率工具可以告诉你，系统哪些部分完全没有被测试覆盖。

coverage.py 是 Python 推荐使用的覆盖率工具。接下来介绍如何在 Tasks 项目中让它与 pytest 一起工作。

在使用 coverage.py 之前，请先安装 pytest-cov 插件，它允许你在 pytest

---

[1] https://docs.python.org/3/library/pdb.html

里使用 coverage.py，并且提供了更多的 pytest 选项。因为 coverage 是 pytest-cov 的一个依赖包，所以安装 pytest-cov 就足够了，它会自动引入 coverage.py。

```
$ pip install pytest-cov
Collecting pytest-cov
  Using cached pytest_cov-2.5.1-py2.py3-none-any.whl
Collecting coverage>=3.7.1 (from pytest-cov)
  Using cached coverage-4.4.1-cp36-cp36m-macosx_10_10_x86_64.whl
...
Installing collected packages: coverage, pytest-cov
Successfully installed coverage-4.4.1 pytest-cov-2.5.1
```

下面对 Task 项目的版本 2 进行覆盖率测试。如果你还在使用版本 1，请先卸载版本 1，然后安装版本 2。

```
$ pip uninstall tasks
Uninstalling tasks-0.1.0:
  /path/to/venv/bin/tasks
  /path/to/venv/lib/python3.6/site-packages/tasks.egg-link
Proceed (y/n)? y
  Successfully uninstalled tasks-0.1.0
$ cd /path/to/code/ch7/tasks_proj_v2
$ pip install -e .
Obtaining file:///path/to/code/ch7/tasks_proj_v2
...
Installing collected packages: tasks
  Running setup.py develop for tasks
Successfully installed tasks
$ pip list
...
tasks (0.1.1, /path/to/code/ch7/tasks_proj_v2/src)
...
```

现在新版本的 Tasks 项目安装好了，运行后就能得到覆盖率报告。

```
$ cd /path/to/code/ch7/tasks_proj_v2
$ pytest --cov=src
===================== test session starts ======================
plugins: mock-1.6.2, cov-2.5.1
collected 62 items

tests/func/test_add.py ...
tests/func/test_add_variety.py ........................
tests/func/test_add_variety2.py ...........
```

```
tests/func/test_api_exceptions.py .........
tests/func/test_unique_id.py .
tests/unit/test_cli.py .....
tests/unit/test_task.py ....
---------- coverage: platform darwin, python 3.6.2-final-0 -----------
Name                            Stmts   Miss   Cover
----------------------------------------------------
src/tasks/__init__.py               2      0    100%
src/tasks/api.py                   79     22     72%
src/tasks/cli.py                   45     14     69%
src/tasks/config.py                18     12     33%
src/tasks/tasksdb_pymongo.py       74     74      0%
src/tasks/tasksdb_tinydb.py        32      4     88%
----------------------------------------------------
TOTAL                             250    126     50%
================== 62 passed in 0.47 seconds ==================
```

虽然当前目录是 tasks_proj_v2，但我们的源码都放在 src 目录中，因此，使用 --cov=src 选项可以单独计算此目录的覆盖率并生成报告。

可以看到有些文件的覆盖率很低，甚至是 0%。tasksdb_pymongo.py 的覆盖率就是 0%，这是因为我们在版本 2 中关闭了 MongoDB。另外一些文件的覆盖率也很低，这个项目在上线前显然要在这些地方添加更多的测试。

我认为应该提高 api.py 和 tasksdb_tinydb 的测试覆盖率。让我们看看 tasksdb_tinydb.py 遗漏了什么，最好的方法是使用 HTML 报告。

使用 --cov-report=html 选项再次运行会得到一个 HTML 报告。

```
$ pytest --cov=src --cov-report=html
==================== test session starts ======================
plugins: mock-1.6.2, cov-2.5.1
collected 62 items

tests/func/test_add.py ...
tests/func/test_add_variety.py ..........................
tests/func/test_add_variety2.py ............
tests/func/test_api_exceptions.py .........
tests/func/test_unique_id.py .
tests/unit/test_cli.py .....
tests/unit/test_task.py ....

---------- coverage: platform darwin, python 3.6.2-final-0 -----------
Coverage HTML written to dir htmlcov
================== 62 passed in 0.45 seconds ==================
```

你可以在浏览器中打开 /htmlcov/index.html 文件（见图 7.1）。

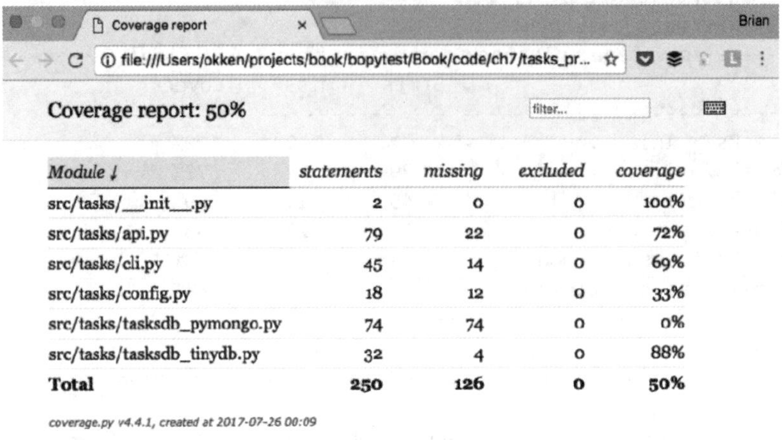

图 7.1 测试覆盖率的 HTML 报告

点击 tasksdb_tinydb.py 会显示该文件的单独报告。报告顶部显示了覆盖率，包括有多少语句被覆盖，有多少语句没有被覆盖（见图 7.2）。

图 7.2 tasksdb_tinydb.py 的单独报告

再往下，你会看到那些未被覆盖的代码（见图 7.3）

从图 7.3 中我们不难看出：

1. 没有测试 list_tasks() 函数里 owner 参数的设置情况。

2. 没有测试 update() 和 delete() 函数。

3. 对 unique_id() 函数的测试不够彻底。

```
31    def list_tasks(self, owner=None):  # type (str) -> list[dict]
32        """Return list of tasks."""
33        if owner is None:
34            return self._db.all()
35        else:
36            return self._db.search(tinydb.Query().owner == owner)
37
38    def count(self):  # type () -> int
39        """Return number of tasks in db."""
40        return len(self._db)
41
42    def update(self, task_id, task):  # type (int, dict) -> ()
43        """Modify task in db with given task_id."""
44        self._db.update(task, eids=[task_id])
45
46    def delete(self, task_id):  # type (int) -> ()
47        """Remove a task from db with given task_id."""
48        self._db.remove(eids=[task_id])
49
50    def delete_all(self):
51        """Remove all tasks from db."""
52        self._db.purge()
53
54    def unique_id(self):  # type () -> int
55        """Return an integer that does not exist in the db."""
56        i = 1
57        while self._db.contains(eids=[i]):
58            i += 1
59        return i
60
```

图 7.3 tasksdb_tinydb.py 中未被覆盖的代码

可以把这些工作列入后面的工作计划里。

虽然测试覆盖率工具非常有用，但是追求 100%的覆盖率是很危险的。因为存在没有被覆盖的代码有可能暗示这部分功能不是必需的，所以可以从系统中删除。像其他软件开发工具一样，测试覆盖率工具并不能替你思考。

coverage.py 和 pytest-cov 还有很多可用的选项，可以在 coverage.py 和 pytest-cov 的文档[1]里查看。

7.3　mock：替换部分系统
mock: Swapping Out Part of the System

mock 包可以用来替换系统的某个部分以隔离要测试的代码。Mock 对象有时被称为 stub、替身。借助 mock 包和 pytest 自身的 monkeypatch（见第 4.5 节），应该可以实现所有的模拟测试。

---

[1] https://coverage.readthedocs.io 和 https://pytest-cov.readthedocs.io

> **对 MOCK 的困惑**
>
> 如果你是第一次使用 mock 对象测试，可能会感到困惑，但实际上它很有趣也很好用。

从 Python 3.3 版本开始，mock 包成为 Python 标准库 unittest.mock 的一部分。在更早的版本中，它是一个单独的 PyPI 安装包。总之，从 Python 2.6 到最新的版本都可以使用 mock 功能。不过，我认为 pytest 的插件 pytest-mock 用起来更方便，我习惯用它来模拟系统。

我们将使用 mock 来测试 Tasks 项目的命令行接口。第 7.2 节曾提到 cli.py 没有被测试覆盖，现在来解决这个问题。

Tasks 项目通过 api.py 文件来测试大多数的功能。因此，命令行测试并不需要完整的功能测试。如果在命令行测试中模拟 API 层，那么我们会很有信心系统可以正常工作。这也方便了在本节中查看 mock 对象。

Tasks 命令行 CLI 的实现使用了第三方接口包 Click[1]。实现 CLI 的方式有很多，包括使用 Python 自带的 argparse 模块。我选用 Click 的原因是它的测试 runner 模块可以帮助我们测试 Click 应用程序。不过 cli.py 里的代码并不是那么明显的 Click 应用程序。

让我们先安装 Tasks 的第 3 版。

```
$ cd /path/to/code/
$ pip install -e ch7/tasks_proj_v2
...
Successfully installed tasks
```

接下来要编写针对 list 功能的一些测试用例。下面先看具体要测试哪些功能：

```
$ tasks list
  ID     owner  done  summary
  --     -----  ----  -------
$ tasks add 'do something great'
```

---

[1] http://click.pocoo.org

```
$ tasks add "repeat" -o Brian
$ tasks add "again and again" --owner Okken
$ tasks list
  ID     owner  done  summary
  --     -----  ----  -------
  1             False do something great
  2      Brian  False repeat
  3      Okken  False again and again
$ tasks list -o Brian
  ID     owner  done  summary
  --     -----  ----  -------
  2      Brian  False repeat
$ tasks list --owner Brian
  ID     owner done summary
  --     ----- ---- -------
  2      Brian False repeat
```

看起来不难。tasks list 命令列出了所有的任务以及标题。即使没有任务，它也会列出标题。如果使用-o 或--owner 参数，那么它将列出该用户的所有任务。该如何测试呢？有很多方法，我们用 mock 来测试。

使用 mock 的测试基本上属于白盒测试，我们必须查看代码，决定要模拟哪些代码。主入口在这里：

ch7/tasks_proj_v2/src/tasks/cli.py
```python
if __name__ == '__main__':
    tasks_cli()
```

下面是对 tasks_cli()函数的调用。

ch7/tasks_proj_v2/src/tasks/cli.py
```python
@click.group(context_settings={'help_option_names': ['-h',
'--help']})
@click.version_option(version='0.1.1')
def tasks_cli():
    """Run the tasks application."""
    pass
```

是不是使 CLI 的代码变得更简洁了？下面是其中的 list 命令：

ch7/tasks_proj_v2/src/tasks/cli.py
```python
@tasks_cli.command(name="list", help="list tasks")
@click.option('-o', '--owner', default=None,
              help='list tasks with this owner')
def list_tasks(owner):
```

```
"""
List tasks in db.

If owner given, only list tasks with that owner.
"""
formatstr = "{: >4} {: >10} {: >5} {}"
print(formatstr.format('ID', 'owner', 'done', 'summary'))
print(formatstr.format('--', '-----', '----', '-------'))
with _tasks_db():
    for t in tasks.list_tasks(owner):
        done = 'True' if t.done else 'False'
        owner = '' if t.owner is None else t.owner
        print(formatstr.format(
            t.id, owner, done, t.summary))
```

习惯 Click 框架下的代码之后，你会发现其实没那么难。我不准备解释所有的代码，毕竟开发命令行代码不是本书的重点。但是，即使我很有信心，也难保我写的代码不会出错，所以好的自动化测试对于保持代码的正确性是非常重要的。

list_tasks(owner)函数依赖其他几个函数：tasks_db()，它是上下文管理器；tasks.list_tasks(owner)是 API 功能函数。我们将使用 mock 来模拟 tasks_db()和 tasks.list_tasks()函数。然后从命令行调用 list_tasks 方法，以确保它正确地调用了 tasks.list_tasks()函数，并且得到正确的返回值。

模拟 task_db()函数，先要看它是如何实现的。

ch7/tasks_proj_v2/src/tasks/cli.py
```
@contextmanager
def _tasks_db():
    config = tasks.config.get_config()
    tasks.start_tasks_db(config.db_path, config.db_type)
    yield
    tasks.stop_tasks_db()
```

task_db()函数是一个上下文管理器，它从 tasks.config.get_config()得到配置信息并借助这些信息建立数据库连接，这又是另一个外部依赖。yield 命令将控制权转移给 list_tasks()函数中的 with 代码块，所有工作完成后会断开数据库连接。

从测试 CLI 命令行调用 API 功能的角度看，我们并不需要一个实际的数据库连接。因此，可以使用一个简单的 stub 来替换上下文管理器。

```
ch7/tasks_proj_v2/tests/unit/test_cli.py
@contextmanager
def stub_tasks_db():
    yield
```

之前没有介绍过 `test_cli.py` 的测试代码，先来看所有的 `import` 语句。

```
ch7/tasks_proj_v2/tests/unit/test_cli.py
from click.testing import CliRunner
from contextlib import contextmanager
import pytest
from tasks.api import Task
import tasks.cli
import tasks.config

@contextmanager
def stub_tasks_db():
    yield
```

这些 `import` 语句是为测试引入的，只有一条 `import` 语句是为 stub 引入的，即 `from contextlib import contextmanager`。

我们将使用 stub 替换真实的上下文管理器。我们会使用到 mocker，它是 pytest-mock 插件提供的 fixture。下面来看一个实际的例子，这个测试用例调用了 `tasks list`。

```
ch7/tasks_proj_v2/tests/unit/test_cli.py
def test_list_no_args(mocker):
    mocker.patch.object(tasks.cli, '_tasks_db', new=stub_tasks_db)
    mocker.patch.object(tasks.cli.tasks, 'list_tasks',
return_value=[])
    runner = CliRunner()
    runner.invoke(tasks.cli.tasks_cli, ['list'])
    tasks.cli.tasks.list_tasks.assert_called_once_with(None)
```

pytest-mock 提供的 mocker 是非常方便的 unittest.mock 接口。第一行代码 `mocker.patch.object(tasks.cli, '_tasks_db', new=stub_tasks_db)` 使用我们的 stub 替换了原来 `tasks_db()` 函数里的上下文管理器。

第二行代码 mocker.patch.object(tasks.cli.tasks,'list_tasks', return_value=[])使用默认的 MagicMock 对象替换了对 tasks.list_task()的调用，然后返回一个空的列表。后面可以用这个对象来检查它是否被正确调用。MagicMock 类是 unittet.Mock 的子类，它可以指定返回值，正如在这个例子中使用的那样。Mock 和 MagicMock 类（包括其他类）使用内置的内省方法模拟其他代码的接口，这让你可以了解它们是如何被调用的。

第三行和第四行代码使用了 Click 框架的 CliRunner 调用 tasks list，就像在命令行调用一样。

最后一行代码使用 mock 对象来确保 API 被正确调用。assert_called_once_with()方法属于 unittest.mock.Mock 对象，完整的方法列表可以参考 Python 文档[1]。

让我们来看一个类似的测试，它用来检查程序的输出。

ch7/tasks_proj_v2/tests/unit/test_cli.py
```
@pytest.fixture()
def no_db(mocker):
    mocker.patch.object(tasks.cli, '_tasks_db', new=stub_tasks_db)

def test_list_print_empty(no_db, mocker):
    mocker.patch.object(tasks.cli.tasks, 'list_tasks',
return_value=[])
    runner = CliRunner()
    result = runner.invoke(tasks.cli.tasks_cli, ['list'])
    expected_output = ("  ID  owner  done  summary\n"
                       "  --  -----  ----  -------\n")

    assert result.output == expected_output
```

这一次把模拟的 tasks_db 放入 no_db fixture 以便我们以后可以很容易复用。tasks.list_tasks()的模拟和之前的例子是一样的。但是这一次我们检查了命令行的输出结果，辨别 result.output 和 expected_output 是否相同。

assert 断言可以放在第一个测试用例 test_list_no_args 里，这样就不需

---
[1] https://docs.python.org/dev/library/unittest.mock.html

要两个测试用例了。但我对自己的 CLI 代码没有什么信心，所以我把代码拆分成了两块："是不是正确地调用了 API？"以及"是不是输出了正确的结果"。用两个测试用例是很恰当的。

其他测试 tasks list 功能的测试用例没有什么特别之处，但是可以帮助我们理解代码。

```python
ch7/tasks_proj_v2/tests/unit/test_cli.py
def test_list_print_many_items(no_db, mocker):
    many_tasks = (
        Task('write chapter', 'Brian', True, 1),
        Task('edit chapter', 'Katie', False, 2),
        Task('modify chapter', 'Brian', False, 3),
        Task('finalize chapter', 'Katie', False, 4),
    )
    mocker.patch.object(tasks.cli.tasks, 'list_tasks',
                        return_value=many_tasks)
    runner = CliRunner()
    result = runner.invoke(tasks.cli.tasks_cli, ['list'])
    expected_output = (" ID   owner done summary\n"
                      " --  -----  ----  -------\n"
                      "  1   Brian  True   write chapter\n"
                      "  2   Katie  False  edit chapter\n"
                      "  3   Brian  False  modify chapter\n"
                      "  4   Katie  False  finalize chapter\n")
    assert result.output == expected_output

def test_list_dash_o(no_db, mocker):
    mocker.patch.object(tasks.cli.tasks, 'list_tasks')
    runner = CliRunner()
    runner.invoke(tasks.cli.tasks_cli, ['list', '-o', 'brian'])
    tasks.cli.tasks.list_tasks.assert_called_once_with('brian')


def test_list_dash_dash_owner(no_db, mocker):
    mocker.patch.object(tasks.cli.tasks, 'list_tasks')
    runner = CliRunner()
    runner.invoke(tasks.cli.tasks_cli, ['list', '--owner', 'okken'])
    tasks.cli.tasks.list_tasks.assert_called_once_with('okken')
```

让我们看看测试结果，以确保它们都正常运行。

```
$ cd /path/to/code/ch7/tasks_proj_v2
$ pytest -v tests/unit/test_cli.py
=================== test session starts ===================
plugins: mock-1.6.2, cov-2.5.1
```

```
collected 5 items

tests/unit/test_cli.py::test_list_no_args PASSED
tests/unit/test_cli.py::test_list_print_empty PASSED
tests/unit/test_cli.py::test_list_print_many_items PASSED
tests/unit/test_cli.py::test_list_dash_o PASSED
tests/unit/test_cli.py::test_list_dash_dash_owner PASSED
================ 5 passed in 0.06 seconds =================
```

太棒了！所有测试都通过。

以上是关于 mock 的简单介绍。如果你想在测试中使用 mock，那么请阅读 unittest.mock 的标准库文档[1]和 pypi.python.org 网站上的 pytest-mock 文档[2]。

7.4 tox：测试多种配置
tox: Testing Multiple Configurations

tox 是一个命令行工具，它允许测试在多种环境下运行。我们将会利用它在多个 Python 版本下测试 Tasks 项目。tox 不仅能测试不同的 Python 版本，还可以用它来测试不同的依赖配置和不同的操作系统配置。

tox 的工作方式大致是这样的：它通过 setup.py 文件为待测程序创建源码安装包；它会查看 tox.ini 中的所有环境设置，并针对每个环境设置执行如下操作。

1. 在 .tox 目录下创建一个虚拟环境。

2. 使用 pip 安装依赖包。

3. 使用 pip 在步骤 1 的虚拟环境中安装自己的程序包。

4. 运行测试用例。

所有环境都测试完后，tox 生成一个汇总的测试结果。

---

[1] https://docs.python.org/dev/library/unittest.mock.html
[2] https://pypi.python.org/pypi/pytest-mock

在实际使用中，观察这些步骤会更好理解，接下来将使用 tox 分别在 Python 2.7 和 Python 3.6 环境下测试 Tasks 项目。选择 Python 2.7 和 Python 3.6 是因为我的计算机上安装了这两个版本。如果你安装的是其他版本，也可以修改 envlist 那一行来指定你的版本。

要对 Tasks 项目做的第一件事是，在项目文件目录最顶层加入 tox.ini 文件（与 setup.py 放在一个目录里）。我还会把 pyteset.ini 的所有内容移植到 tox.ini 文件里。

这是代码目录的大致结构：

```
tasks_proj_v2/
├── ...
├── setup.py
├── tox.ini
├── src
│   └── tasks
│       ├── __init__.py
│       ├── api.py
│       └── ...
└── tests
    ├── conftest.py
    ├── func
    │   ├── __init__.py
    │   ├── test_add.py
    │   └── ...
    └── unit
        ├── __init__.py
        ├── test_task.py
        └── ...
```

现在我们来看 tox.ini 文件的内容。

`ch7/tasks_proj_v2/tox.ini`
```
# tox.ini , put in same dir as setup.py

[tox]
envlist = py27,py36

[testenv]
deps=pytest
commands=pytest
```

```
[pytest]
addopts = -rsxX -l --tb=short --strict
markers =
    smoke: Run the smoke test test functions
    get: Run the test functions that test tasks.get()
```

在[tox]标签下，我们指定 envlist = py27,py36，代表 tox 使用 python2.6 和 python3.6 来运行我们的测试用例。

在[testenv]标签下，deps=pytest 这一行告诉 tox 确保 pytest 已经安装。如果有多个测试依赖，则可以按行分别罗列出来，同时还可以指定版本。

commands=pytest 这一行用来指示 tox 在每个环境里运行 pytest。

在[pytest]标签下，我们可以把平常放在 pytest.ini 文件里的配置放在这里。本例中，addopts 用来显示跳过的、预计失败的、预计通过（-rsxX）的测试信息，并且显示堆栈中的局部变量（-l）。同时指定使用精简的回溯信息（--tb=short）并且确保标记都是严格声明过的（--strict）。

在运行 tox 之前，请确认 tox 已安装。

```
$ pip install tox
```

这个命令可以在虚拟环境里执行。

接下来运行 tox，只需要输入 tox 即可。

```
$ cd /path/to/code/ch7/tasks_proj_v2
$ tox
GLOB sdist-make: /path/to/code/ch7/tasks_proj_v2/setup.py
py27 create: /path/to/code/ch7/tasks_proj_v2/.tox/py27
py27 installdeps: pytest
py27 inst: /path/to/code/ch7/tasks_proj_v2/.tox/dist/tasks-0.1.1.zip
py27 installed: click==6.7,funcsigs==1.0.2,mock==2.0.0,
                pbr==3.1.1,py==1.4.34,pytest==3.2.1,
pytest-mock==1.6.2,six==1.10.0,tasks==0.1.1,tinydb==3.4.0
py27 runtests: PYTHONHASHSEED='1311894089'
py27 runtests: commands[0] | pytest
================= test session starts =================
plugins: mock-1.6.2
```

```
collected 62 items

tests/func/test_add.py ...
tests/func/test_add_variety.py ..........................
tests/func/test_add_variety2.py ............
tests/func/test_api_exceptions.py .........
tests/func/test_unique_id.py .
tests/unit/test_cli.py .....
tests/unit/test_task.py ....
============== 62 passed in 0.25 seconds ===============
py36 create: /path/to/code/ch7/tasks_proj_v2/.tox/py36
py36 installdeps: pytest
py36 inst: /path/to/code/ch7/tasks_proj_v2/.tox/dist/tasks-0.1.1.zip
py36 installed: click==6.7,py==1.4.34,pytest==3.2.1,
pytest-mock==1.6.2,six==1.10.0,tasks==0.1.1,tinydb==3.4.0
py36 runtests: PYTHONHASHSEED='1311894089'
py36 runtests: commands[0] | pytest
================ test session starts =================
plugins: mock-1.6.2
collected 62 items

tests/func/test_add.py ...
tests/func/test_add_variety.py ..........................
tests/func/test_add_variety2.py ............
tests/func/test_api_exceptions.py .........
tests/func/test_unique_id.py .
tests/unit/test_task.py ....
============== 62 passed in 0.27 seconds ===============
_____ summary _____
  py27: commands succeeded
  py36: commands succeeded
  congratulations :)
```

最后得到了所有测试环境下的汇总报告。

```
_____ summary _____
  py27: commands succeeded
  py36: commands succeeded
  congratulations :)
```

收到恭喜和笑脸，是不是很愉快呢？

tox 的功能远比我这里展示的更强大。如果你经常使用 pytest 在多个环境下测试，它能帮上大忙。请查阅 tox 文档了解更详细的信息。

7.5 Jenkins CI:让测试自动化
Jenkins CI: Automating Your Automated Tests

Jenkins 之类的 CI(持续集成)系统常用来在提交代码后启动测试组件。pytest 可以生成 junit.xml 格式的文件,用于给 Jenkins 和其他 CI 系统显示测试结果。

Jenkins 是一个开源的自动化系统持续集成服务器。虽然 Python 不需要编译,但许多人还是习惯用 Jenkins 与其他 CI 系统来自动化运行和测试 Python 项目。本节将介绍如何在 Jenkins 中设置 Tasks 项目。我不打算介绍 Jenkins 的安装过程,不同的操作系统安装 Jenkins 有不同的要求,请查阅 Jenkins 网站上的说明[1]。

使用 Jenkins 运行 pytest 测试组件,有几个 Jenkins 插件特别好用。为了测试示例项目,我安装了以下插件。

- build-name-setter:这个插件可以设置每次运行的名字,而不是默认的 #1、#2、#3。

- Test Result Analyzer plugin:这个插件以图表的形式显示测试执行的历史信息。

你可以在 Jenkins 的主页上安装插件。本机访问 local-host:8080/manage,然后依次点击 Manage Jenkins→Manage Plugins→Available。在过滤框中可以搜索你想要的插件,然后勾选。我通常会选择"Install without Restart",然后在安装/升级插件页面勾选"Restart Jenkins when Installation is complete and no jobs are running"。

让我们看一个完整的配置过程。在 Jenkins 的"project/item"选项中选择"Freestyle Project",命名为 tasks,如图 7.4 所示。

---

[1] https://jenkins.io

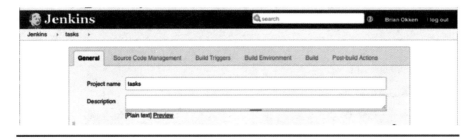

图 7.4　Jenkins 的配置页面

这个配置有些奇怪，因为我们要使用 Tasks 项目的不同版本，如 tasks_proj、tasks_proj_v2 等，而不是使用版本控制。因此，要使用参数来指示每个测试会话在哪里安装 Tasks 项目，以及在哪里寻找测试用例。我们将用几个字符串参数来指示这些路径。点击"This project is parameterized"显示可用的选项，如图 7.5 所示。

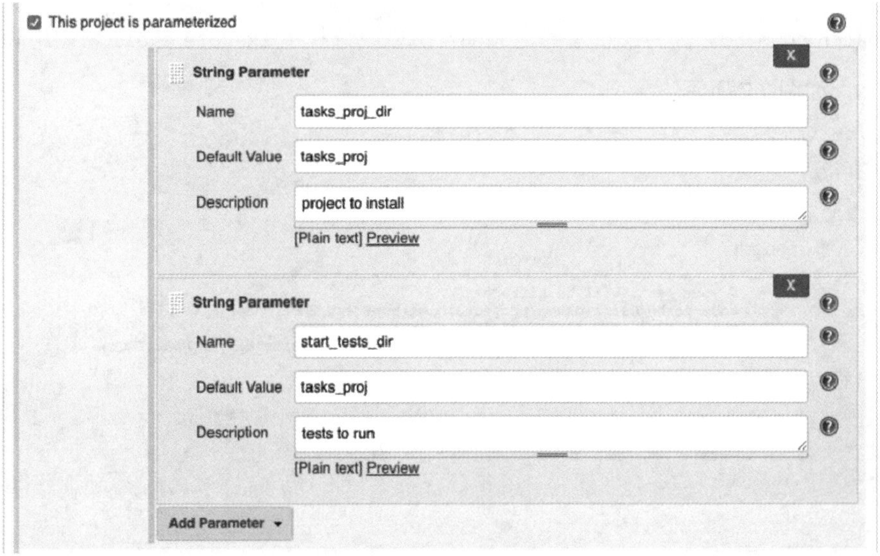

图 7.5　设置字符串参数

然后，请向下找到"Build Environment"，选择"Delete workspace before build starts"，然后选择"Set Build Name"，将名字设置为

${start_tests_dir}#${BUILD_NUMBER}，如图 7.6 所示。

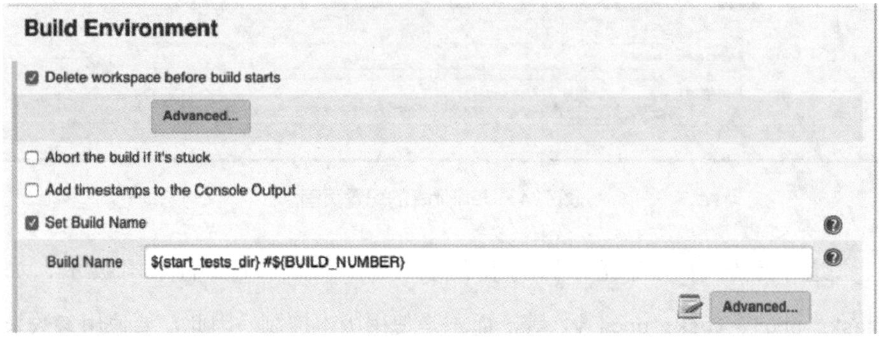

图 7.6 设置 Build Environment

接下来是构建。在 Mac 或 UNIX 这类系统上，选择 Add build step -> Execute shell。在 Windows 上，选择 Add build step -> Execute Windows batch command。因为我使用的是 Mac，所以我选择了 Execute shell 来调用脚本，如图 7.7 所示。

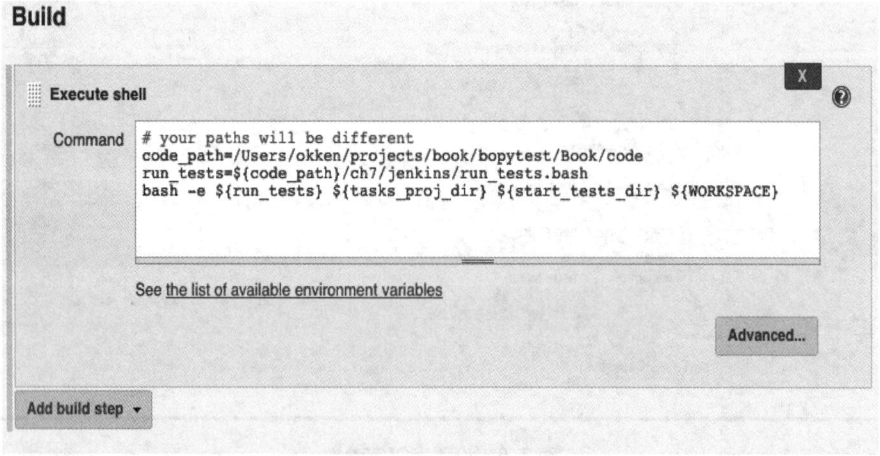

图 7.7 设置脚本

7.5 Jenkins CI：让测试自动化

图 7.7 中的代码如下：

```
# your paths will be different
code_path=/Users/okken/projects/book/bopytest/Book/code
run_tests=${code_path}/ch7/jenkins/run_tests.bash
bash -e ${run_tests} ${tasks_proj_dir} ${start_tests_dir} ${WORKSPACE}
```

我使用了一个脚本，而不是把所有代码都放在 Jenkins 的执行框里，这样所有的修改都可以由版本控制记录下来。脚本如下：

```
ch7/jenkins/run_tests.bash
#!/bin/bash

# your paths will be different
top_path=/Users/okken/projects/book/bopytest/Book
code_path=${top_path}/code
venv_path=${top_path}/venv
tasks_proj_dir=${code_path}/$1
start_tests_dir=${code_path}/$2
results_dir=$3

# click and Python 3,
# from http://click.pocoo.org/5/python3/
export LC_ALL=en_US.utf-8
export LANG=en_US.utf-8

# virtual environment
source ${venv_path}/bin/activate

# install project
pip install -e ${tasks_proj_dir}

# run tests
cd ${start_tests_dir}
pytest --junit-xml=${results_dir}/results.xml
```

最后一行代码是 pytest --junit-xml=${results_dir}/results.xml，其中的--junit-xml 选项是 Jenkins 用来生成 junit.xml 结果文件所需的标识。

junit 还有其他选项：

```
$ pytest --help | grep junit
  --junit-xml=path create junit-xml style report file at given path.
  --junit-prefix=str prepend prefix to classnames in junit-xml output
  junit_suite_name (string) Test suite name for JUnit report
```

--junit-prefix 选项可以作为前缀用在每一个测试用例里。区分不同环境下的测试结果时，它很有用。junit_suite_name 是一个配置文件选项，可以在 pytest.ini 或 tox.ini 文件的[pytest]区域设置。稍后我们会看到结果中包含了 from(pytest)，可以使用 junit_suite_name 把 pytest 变成你想要的名字。

接下来找到 Post-build Actions -> Publish Junit test result report，在 Test report XMLs 栏填入 results.xml，如图 7.8 所示。

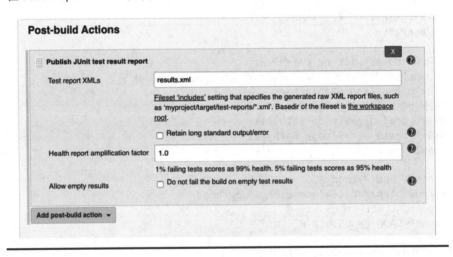

图 7.8　设置测试结果报告

现在可以使用 Jenkins 来进行测试了，步骤如下。

1. 点击 Save。

2. 来到项目顶层。

3. 点击"Build with Parameters"。

4. 选择你的文件目录并点击 Build。

5. 完成后，移动鼠标到 Build History 的名字上，点击悬浮菜单里的 Console Output（或者点击任务名字后选择 Console Output）。

6. 查看输出结果，如果出错了，请思考问题出在哪里。

如果你运气好的话，可能会略过第 5 步和第 6 步，但是我从来没有这样的好运气。设置完 Jenkins 任务后，总是会遇到各种问题（目录权限、路径、脚本中存在拼写错误，等等）。

查看结果之前可以多运行几次，使图表呈现的结果更有对比性。再次点击"Build with Parameters"，选择同样的项目目录，但是把 start_test_dir 设置为 ch2，然后点击 Build。刷新项目视图后，你会看到如图 7.9 所示的图形。

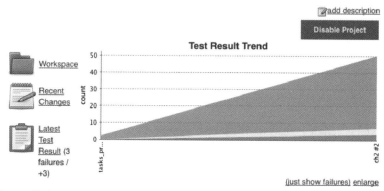

图 7.9　刷新后的项目视图

点击"Latest Test Result"可以查看测试会话的摘要，点击图标+可以展示失败的用例。

点击任意失败的测试用例名字会展示失败信息（见图 7.10）。这里你看到了 (from pytest) 字符串，它是测试名字的一部分，由配置文件里的 junit_suite_name 选项控制。

```
Failed
tasks_proj.tests.func.test_unique_id_1.test_unique_id (from pytest)

                                    Failing for the past 1 build (Since #2)
                                                              Took 2 ms.
                                                           add description

Error Message
assert 1 != 1

Stacktrace
def test_unique_id():
        id_1 = tasks.unique_id()
        id_2 = tasks.unique_id()
>       assert id_1 != id_2
E       assert 1 != 1

tasks_proj/tests/func/test_unique_id_1.py:9: AssertionError
```

图 7.10 失败信息

返回到 Jenkins→tasks，点击 Test Results Analyzer 可以查看各个测试会话里有哪些测试用例没有执行，有哪些通过了，有哪些失败了（见图 7.11）。

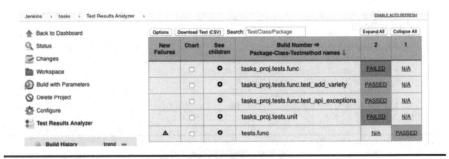

图 7.11 查看测试结果

你已经看到在 Jenkins 里如何利用虚拟环境运行 pytest 测试组件，但是 Jenkins 的用途还远不止于此，比如针对多个环境的测试，不仅可以在各个环境设置不同的 Jenkins 任务，还可以让 Jenkins 直接调用 tox。有一个很棒的插件 Cobertura，它可以显示 coverage.py 的覆盖率数据（请查阅 Jenkins 文档）。

7.6 unittest：用 pytest 运行历史遗留测试用例
unittest: Running Legacy Tests with pytest

unittest 是 Python 内置的标准测试库。它原本是用来测试 Python 自身的，但也经常用于项目测试。pytest 可以执行 unittest，而且可以在一个测试会话里同时执行 pytest 和 unittest。

假设 Tasks 项目开始是用 unittest 做测试，而不是用 pytest，而且已经写了很多测试用例。你可以使用 pytest 来运行基于 unittest 的测试用例。如果想从 unittest 过渡到 pytest，当然很好。但也可以保留所有的 unittest 测试用例，同时使用 pytest 编写新的测试用例。如果时间允许，你可以慢慢迁移老的测试用例。在迁移过程中或许会遇到一些问题，稍后会谈到。下面看看 unittest 下的一个测试用例：

```
ch7/unittest/test_delete_unittest.py
import unittest
import shutil
import tempfile
import tasks
from tasks import Task

def setUpModule():
    """Make temp dir, initialize DB."""
    global temp_dir
    temp_dir = tempfile.mkdtemp()
    tasks.start_tasks_db(str(temp_dir), 'tiny')

def tearDownModule():
    """Clean up DB, remove temp dir."""
    tasks.stop_tasks_db()
    shutil.rmtree(temp_dir)

class TestNonEmpty(unittest.TestCase):

    def setUp(self):
        tasks.delete_all()  # start empty
        # add a few items, saving ids
        self.ids = []
        self.ids.append(tasks.add(Task('One', 'Brian', True)))
        self.ids.append(tasks.add(Task('Two', 'Still Brian', False)))
        self.ids.append(tasks.add(Task('Three', 'Not Brian', False)))
```

```python
    def test_delete_decreases_count(self):
        # GIVEN 3 items
        self.assertEqual(tasks.count(), 3)
        # WHEN we delete one
        tasks.delete(self.ids[0])
        # THEN count decreases by 1
        self.assertEqual(tasks.count(), 2)
```

真正的测试在最下面,test_delete_decreases_count()。其余代码都是用来设置和清理测试环境的。这个测试在 unittest 下可以正常运行。

```
$ cd /path/to/code/ch7/unittest
$ python -m unittest -v test_delete_unittest.py
test_delete_decreases_count (test_delete_unittest.TestNonEmpty) ... ok

----------------------------------------------------------------------
Ran 1 test in 0.024s
OK
```

在 pytest 下也运行正常。

```
$ pytest -v test_delete_unittest.py
=========================== test session starts ============================
collected 1 items

test_delete_unittest.py::TestNonEmpty::test_delete_decreases_count PASSED
========================= 1 passed in 0.02 seconds =========================
```

如果只想用 pytest 来运行 unittest,这没有问题。但是我们的初衷是把 Tasks 项目迁移到 pytest。假设我们希望一次迁移一个测试,并且同时运行 pytest 和 unittest 版本的测试用例,直到得到完全满意的 pytest 版本。下面是重写后的测试,我们来同时运行它们。

ch7/unittest/test_delete_pytest.py
```python
import tasks


def test_delete_decreases_count(db_with_3_tasks):
    ids = [t.id for t in tasks.list_tasks()]
    # GIVEN 3 items
    assert tasks.count() == 3
    # WHEN we delete one
    tasks.delete(ids[0])
```

```
    # THEN count decreases by 1
    assert tasks.count() == 2
```

Tasks 项目使用的 fixture（如第 3.5 节介绍过的 db_with_3_tasks）可以在测试前设置好数据库。使用 pytest 重构之后代码量变小了，但测试功能是一模一样的。

两个测试都通过了。

```
$ pytest -q test_delete_pytest.py
.1
passed in 0.01 seconds
$ pytest -q test_delete_unittest.py
.1
passed in 0.02 seconds
```

我们甚至可以同时运行它们。注意要让 unittest 的版本首先运行。

```
$ pytest -v test_delete_unittest.py test_delete_pytest.py
=========================== test session starts ============================
collected 2 items

test_delete_unittest.py::TestNonEmpty::test_delete_decreases_count PASSED
test_delete_pytest.py::test_delete_decreases_count[tiny] PASSED
========================= 2 passed in 0.07 seconds =========================
```

如果先运行 pytest 的版本，就会出错。

```
$ pytest -v test_delete_pytest.py test_delete_unittest.py
=========================== test session starts ============================
collected 2 items

test_delete_pytest.py::test_delete_decreases_count[tiny] PASSED
test_delete_unittest.py::TestNonEmpty::test_delete_decreases_count PASSED
test_delete_unittest.py::TestNonEmpty::test_delete_decreases_count ERROR
================================== ERRORS ==================================
_____ ERROR at teardown of TestNonEmpty.test_delete_decreases_count _____

tmpdir_factory = <_pytest.tmpdir.TempdirFactory object at 0x1038a3128>
request = <SubRequest 'tasks_db_session'
```

```
                  for <Function 'test_delete_decreases_count[tiny]'>>

    @pytest.fixture(scope='session', params=['tiny'])
    def tasks_db_session(tmpdir_factory, request):
        temp_dir = tmpdir_factory.mktemp('temp')
        tasks.start_tasks_db(str(temp_dir), request.param)
        yield # this is where the testing happens
>       tasks.stop_tasks_db()

conftest.py:11:
_ _ _ _ _ _ _ _ _ _ _ _ _ _ _ _ _ _ _ _ _ _ _ _ _ _ _ _ _ _ _ _ _ _

    def stop_tasks_db(): # type: () -> None
        global _tasksdb
>       _tasksdb.stop_tasks_db()
E       AttributeError: 'NoneType' object has no attribute
'stop_tasks_db'

../tasks_proj_v2/src/tasks/api.py:104: AttributeError
===================== 2 passed, 1 error in 0.13 seconds
=====================
```

可以看到两个测试用例都通过了,但是在最后关闭数据库时出现了异常。

让我们使用--setup-show选项来找找原因。

```
$ pytest -q --tb=no --setup-show test_delete_pytest.py
test_delete_unittest.py

SETUP    S tmpdir_factory
SETUP    S tasks_db_session (fixtures used: tmpdir_factory)[tiny]
    SETUP    F tasks_db (fixtures used: tasks_db_session)
SETUP    S tasks_just_a_few
    SETUP    F db_with_3_tasks (fixtures used: tasks_db,
tasks_just_a_few)
        test_delete_pytest.py::test_delete_decreases_count[tiny]
           (fixtures used: db_with_3_tasks, tasks_db, tasks_db_session,
                          tasks_just_a_few, tmpdir_factory).
        TEARDOWN F db_with_3_tasks
        TEARDOWN F tasks_db

test_delete_unittest.py::TestNonEmpty::test_delete_decreases_count.
TEARDOWN S tasks_just_a_few
TEARDOWN S tasks_db_session[tiny]
TEARDOWN S tmpdir_factoryE
2 passed, 1 error in 0.08 seconds
```

清理测试环境的fixture是在所有测试用例(包括unittest的测试用例)运行

完后执行的。这让我困惑了一阵子，直到我意识到 unittest 模块里的 tearDownModule()关闭了数据库连接。而 pytest 里的 tasks_db_session()函数后来又试图做同样的事情。

解决办法是针对 unittest 的测试用例，使用 pytest 会话范围的 fixture。这可以通过在类或方法层面添加@pytest.mark.usefixtures()装饰器来实现。

ch7/unittest/test_delete_unittest_fix.py
```python
import pytest
import unittest
import tasks
from tasks import Task

@pytest.mark.usefixtures('tasks_db_session')
class TestNonEmpty(unittest.TestCase):

    def setUp(self):
        tasks.delete_all() # start empty
        # add a few items, saving ids
        self.ids = []
        self.ids.append(tasks.add(Task('One', 'Brian', True)))
        self.ids.append(tasks.add(Task('Two', 'Still Brian', False)))
        self.ids.append(tasks.add(Task('Three', 'Not Brian', False)))

    def test_delete_decreases_count(self):
        # GIVEN 3 items
        self.assertEqual(tasks.count(), 3)
        # WHEN we delete one
        tasks.delete(self.ids[0])
        # THEN count decreases by 1
        self.assertEqual(tasks.count(), 2)
```

现在 unittest 和 pytest 就可以共存而不会互相干扰了。

```
$ pytest -v test_delete_pytest.py test_delete_unittest_fix.py
==================== test session starts ====================
plugins: mock-1.6.0, cov-2.5.1
collected 2 items

test_delete_pytest.py::test_delete_decreases_count PASSED
test_delete_unittest_fix.py::TestNonEmpty::test_delete_decreases_cou
nt PASSED
================== 2 passed in 0.02 seconds ==================
```

请注意，只有当 unittest 和 pytest 在测试会话中共享资源时才需要这样做。我们在第 2.4 节介绍过测试函数的标记，unitest 的测试用例上也可以使用 pytest 标记，比如 @pytest.mark.skip() 和 @pytest.mark.xfail()，还有用户自定义的标记 @pytest.mark.foo() 等。

回到 unittest 示例，我们依然使用 setUp() 函数来保存所有 tasks 的 ID。显然，得到 tasks 的 ID 列表是一个很常见的 API 方法，但该函数也反映出 pytest.mark.usefixures 和 unittest 一起使用时的一个小问题：我们不能直接从 fixture 给 unittest 函数传递数据。

但是，你可以通过 cls 对象（request 对象的一部分）来传递数据。在下面的例子里，setUp() 函数里的代码被移植到一个作用范围是函数级别的 fixture，它可以通过 request.cls.ids 传递 ids。

ch7/unittest/test_delete_unittest_fix2.py
```python
import pytest
import unittest
import tasks
from tasks import Task

@pytest.fixture()
def tasks_db_non_empty(tasks_db_session, request):
    tasks.delete_all() # start empty
    # add a few items, saving ids
    ids = []
    ids.append(tasks.add(Task('One', 'Brian', True)))
    ids.append(tasks.add(Task('Two', 'Still Brian', False)))
    ids.append(tasks.add(Task('Three', 'Not Brian', False)))
    request.cls.ids = ids

@pytest.mark.usefixtures('tasks_db_non_empty')
class TestNonEmpty(unittest.TestCase):

    def test_delete_decreases_count(self):
        # GIVEN 3 items
        self.assertEqual(tasks.count(), 3)
        # WHEN we delete one
        tasks.delete(self.ids[0])
        # THEN count decreases by 1
        self.assertEqual(tasks.count(), 2)
```

这个测试用例通过 `self.ids` 访问 `ids` 列表，与之前一样。

使用标记有一个限制：基于 unittest 的测试用例不能使用参数化的 fixture。不过，最后一个例子同时使用了 pytest fixture 和 unittest，把它重构成 pytest 格式的测试用例并不难，只需要去掉 `unittest.TestCase` 基类并且用 `assert` 调用替换 `self.assertEqual()` 函数就可。

用 pytest 运行 unittest 还有一个问题：unittest 的测试子集会在首次遇错时停止执行，但是单独使用 unittest 时，无论是否有错，unittest 都会依次运行每个测试子集。除非所有的测试子集都能通过，否则 pytest 不会全部执行。但我觉得这是一个小问题。

7.7 练习
Exercises

1. 第 2 章的测试代码中有一些刻意令它们失败的测试用例。请使用`--pdb`选项来运行这些测试用例。不要使用 -x 选项，这样每次失败时，调试器就会打开。

2. 请尝试修复这些代码并使用`--lf --pdb` 选项和调试器重新运行这些测试用例。最好在没有项目压力的情况下练习使用这些调试工具。

3. 在统计测试覆盖率时，注意到很多代码没有覆盖到，`tasks.update()` 是其中之一。请在 `func` 目录下添加一些用于测试 `tasks.update()`的测试用例。

4. 运行 `coverage.py`，查看还有哪些代码没有被测试覆盖。思考是否 `api.py` 完全覆盖，整个项目就被完全覆盖。

5. 在 `test_cli.py` 文件里添加几个测试用例来检查 `tasks update` 命令行接口，请使用 `mock`。

6. 使用 `tox` 在至少两个 Python 版本里运行新测试用例（与老测试用例一

起运行）。

7. 使用 Jenkins 对不同版本的 `tasks_proj` 代码和测试用例进行组合配置，并生成图像。

7.8 预告
What's Next

我想读者现在已经掌握了在项目中使用 pytest 的要领了。接下来是附录，附录的内容也很值得一读，比如附录 C（讲解插件包）和附录 D（讲解如何打包以及共享代码）。附录 E 介绍了另一种形式的 fixture，它与传统的 xUnit 测试工具更像。

最后，请读者与我们保持联系。你可以访问本书官网，在讨论区留言并指出书中的错误，帮助我们优化本书的后续版本，使其更简洁，更易理解。我希望本书成为一本活的 pytest 文档。我们会根据 pytest 的官方开发情况，不断地为 pytest 用户提供有价值的内容。

附录 A
虚拟环境
Virtual Environments

Python 虚拟环境是一个沙箱，它使用独立的软件包，而不是使用系统提供的 `site-packages` 文件夹。使用虚拟环境的场景有很多，比如你要在同一个 Python 环境下运行多个服务，但是每个服务使用的软件包和版本都不相同；再比如要为每个 Python 项目配置独立的软件包，虚拟环境就可以派上用场。

从 Python 3.3 版本开始，虚拟环境 venv 模块成为了标准库的一部分，不过在 Ubuntu 环境下，venv 出现过一些问题。而 PyPI 提供的 `virtualenv` 在 Ubuntu 上可以支持 Python 3.6（向后兼容至 2.6）。大多数情况下，`virtualenv` 就足够用了，因此接下来就以 `virtualenv` 来介绍虚拟环境的使用。

在 macOS 和 Linux 系统中，可以运行以下命令设置虚拟环境。

```
$ pip install -U virtualenv
$ virtualenv -p /path/to/a/python.exe /path/to/env_name
$ source /path/to/env_name/bin/activate
(env_name) $
... do your work ...
(env_name) $ deactivate
```

你也可以使用 Python 来执行上述命令。

```
$ python3.6 -m pip install -U virtualenv
$ python3.6 -m virtualenv env_name
$ source env_name/bin/activate
(env_name) $
... do your work ...
(env_name) $ deactivate
```

在 Windows 上，`activate` 那行命令有点不一样。

```
C:/>pip install -U virtualenv
C:/>virtualenv -p /path/to/a/python.exe /path/to/env_name
C:/>/path/to/env_name/Scripts/activate.bat
(env_name) C:/>
... do your work ...
(env_name) C:/> deactivate
```

在 Windows 上，你同样可以使用 Python 来运行上述命令。在实际工作中，启动虚拟环境要求的步骤更少，因为我们不需要频繁地更新 virtualenv。我通常把虚拟环境的目录 env_name 放置在项目目录的最顶层。我的步骤通常如下：

```
$ cd /path/to/my_proj
$ virtualenv -p $(which python3.6) my_proj_venv
$ source my_proj_venv/bin/activate
(my_proj_venv) $
... do your work ...
(my_proj_venv) $ deactivate
```

我也见过另外两种有趣的安装方式。

1. 将虚拟环境放置在项目目录中（像上面的代码那样），统一命名 env 目录，比如 venv 或 .venv。这样做的好处是你可以把 venv 或 .venv 放入全局 .gitignore 文件里。坏处是命令提示符只显示你在使用虚拟环境，但不知道使用的是哪一个。

2. 将所有虚拟环境放入一个共同目录，比如 ~/venvs/。现在环境名字是不一样的，命令提示符会更有意义。你也不需要担心 .gitignore 文件，因为它并不在项目目录里。这个目录也方便你查看有哪些项目。

请记住，一个虚拟环境是一个目录，它可链接 `python.exe` 文件、`pip.exe` 文件和它所使用的 Python 版本。不过你安装的东西只会存在于虚拟环境的目录里，而不会出现在全局的 `site-packages` 目录里。使用完虚拟环境后，你可以将该目录删除，所有的东西也随之消失。

以上只是 virtualenv 最基本的用法。virtualenv 是相当灵活的工具，包含许多个性化的选项。请务必查阅 `virtualenv--help`，其中可能已经包含你会遇到的各种问题。如果还有疑惑，可以查看 https://virtualenv.pypa.io。

附录 B
Pip
Pip

pip 是安装 Python 包的工具，它是 Python 组件之一。有人开玩笑说 pip 是 Pip Installs Python 或者 Pip Installs Packages 的缩写（这里用到了递归定义）。如果你安装了好几个版本的 Python，那么每个版本都有它自己的 pip 程序。

运行 pip install something，默认情况下 pip 将会：

1. 连接到 PyPI 仓库（http://pypi.python.org/pypi）。
2. 查找名为 something 的包。
3. 下载适合你的系统和 Python 版本的 something 包版本。
4. 将 something 包安装到 Python 安装目录下的 site-packages 目录里。

这是 pip 大致的工作流程。pip 还可以用来定制某个包的启动脚本，做 wheel 缓存，等等。

正如上面提到的，每个安装好的 Python 都有它自己的 pip。如果使用虚拟环境，pip 和 python 就会自动链接到创建虚拟环境时指定的 Python 版本。如果没有使用虚拟环境，但安装了多个版本的 Python（如 Python 3.5 和 Python 3.6），则可能需要使用 python3.5 -m pip 或 python3.6 -m pip，而不是直接输入 pip。

当然，它们的效果是一样的（附录中的示例假定读者使用了虚拟环境，并且pip能正常工作）。

要查看pip的版本和它所关联的Python版本，请使用命令pip --version。

```
(my_env) $ pip --version
pip 9.0.1 from /path/to/code/my_env/lib/python3.6/site-packages (python 3.6)
```

要查看所有pip安装的软件包，请使用命令pip list。如果需要卸载某些包，请运行pip uninstall something命令。

```
(my_env) $ pip list
pip (9.0.1)
setuptools (36.2.7)
wheel (0.29.0)
(my_env) $ pip install pytest
...
Installing collected packages: py, pytest
Successfully installed py-1.4.34 pytest-3.2.1
(my_env) $ pip list
pip (9.0.1)
py (1.4.34)
pytest (3.2.1)
setuptools (36.2.7)
wheel (0.29.0)
```

如本例所示，pip安装了所有需要安装的依赖包。

pip很灵活，安装来源也很多，可以从GitHub安装，也可以从私人服务器安装，还可以从共享目录或者本地的开发包安装。除非使用虚拟环境，否则软件包通常会安装到site-packages目录。

只要是在PyPI上完成发布的软件包，pip就可以在pypi.python.org中搜索到，并安装指定版本。

```
$ pip install pytest==3.2.1
```

pip可以安装包含setup.py文件的本地包。

```
$ pip install /path/to/package
```

本地安装的话，如果在包所在的目录，请使用./package_name。

```
$ cd /path/just/above/package
$ pip install my_package # pip 从 PyPI 查找 "my_package"包
$ pip install ./my_package # pip 从本地目录查找包
```

pip 可以直接安装下载好的 zip 包和 wheel 包，而不必对它们进行手动解压。

requirements.txt 可以存储依赖列表，pip 读入后将一次性下载并安装多个依赖包。

```
(my_env) $ cat requirements.txt
pytest==3.2.1
pytest-xdist==1.20.0
(my_env) $ pip install -r requirements.txt
...
Successfully installed apipkg-1.4 execnet-1.4.1 pytest-3.2.1
pytest-xdist-1.20.0
```

你可以使用 pip 把不同版本的包下载到本地，然后把 pip 指向那里（而不是 PyPI），以后即使在离线情况下也能把它们安装到虚拟环境。

以下示例下载了 pytest 和所有的依赖包。

```
(my_env) $ mkdir ~/.pipcache
(my_env) $ pip download -d ~/pipcache pytest
Collecting pytest
  Using cached pytest-3.2.1-py2.py3-none-any.whl
  Saved /Users/okken/pipcache/pytest-3.2.1-py2.py3-none-any.whl
Collecting py>=1.4.33 (from pytest)
  Using cached py-1.4.34-py2.py3-none-any.whl
  Saved /Users/okken/pipcache/py-1.4.34-py2.py3-none-any.whl
Collecting setuptools (from pytest)
  Using cached setuptools-36.2.7-py2.py3-none-any.whl
  Saved /Users/okken/pipcache/setuptools-36.2.7-py2.py3-none-any.whl
Successfully downloaded pytest py setuptools
```

以后即使在离线的情况下，你也可以从本地缓存安装。

```
Collecting pytest
Collecting py>=1.4.33 (from pytest)
...
Installing collected packages: py, pytest
Successfully installed py-1.4.34 pytest-3.2.1
```

这个技巧在使用 tox 和做持续集成时非常有用。旅行之前我通常会用这个办法下载一些依赖包，以便在旅途中（比如在飞机上）编写代码。其他有关 pip 的用法请参考官方文档[1]。

---

[1] https://pip.pypa.io

附录 C

常用插件
Plugin Sampler Pack

插件是 pytest 的重要功能，它可以大大简化测试的工作。好用的插件有很多，比如第 7.2 节介绍过的 `pytest-cov` 插件，以及第 7.3 节介绍过的 `pytest-mock` 插件。这里再介绍几个具有代表性的插件。

本节介绍的所有插件都可以在 PyPI 上找到，并且可以通过命令 `pip install <plugin-name>` 来安装。

C.1 改变测试流程的插件
Plugins That Change the Normal Test Run Flow

接下来介绍的插件可以改变 pytest 执行测试用例的方式。

pytest-repeat：重复运行测试
pytest-repeat: Run Tests More Than Once

如果希望在一个会话中重复运行测试，则可以使用 pytest-repeat 插件[1]。如果测试总是断断续续地失败，那么这个插件就很有用。

---

[1] https://pypi.python.org/pypi/pytest-repeat

下面是 ch7/tasks_proj_v2 里以 test_list 开头的测试用例的正常运行情况。

```
$ cd /path/to/code/ch7/tasks_proj_v2
$ pip install .
$ pip install pytest-repeat
$ pytest -v -k test_list
====================== test session starts ======================
plugins: repeat-0.4.1, mock-1.6.2
collected 62 items

tests/func/test_api_exceptions.py::test_list_raises PASSED
tests/unit/test_cli.py::test_list_no_args PASSED
tests/unit/test_cli.py::test_list_print_empty PASSED
tests/unit/test_cli.py::test_list_print_many_items PASSED
tests/unit/test_cli.py::test_list_dash_o PASSED
tests/unit/test_cli.py::test_list_dash_dash_owner PASSED
====================== 56 tests deselected ======================
=========== 6 passed, 56 deselected in 0.10 seconds ============
```

有了 pytest-repeat 插件，可以使用--count 来指定每个测试用例运行两次。

```
$ pytest --count=2 -v -k test_list
====================== test session starts ======================
plugins: repeat-0.4.1, mock-1.6.2
collected 124 items

tests/func/test_api_exceptions.py::test_list_raises[1/2] PASSED
tests/func/test_api_exceptions.py::test_list_raises[2/2] PASSED
tests/unit/test_cli.py::test_list_no_args[1/2] PASSED
tests/unit/test_cli.py::test_list_no_args[2/2] PASSED
tests/unit/test_cli.py::test_list_print_empty[1/2] PASSED
tests/unit/test_cli.py::test_list_print_empty[2/2] PASSED
tests/unit/test_cli.py::test_list_print_many_items[1/2] PASSED
tests/unit/test_cli.py::test_list_print_many_items[2/2] PASSED
tests/unit/test_cli.py::test_list_dash_o[1/2] PASSED
tests/unit/test_cli.py::test_list_dash_o[2/2] PASSED
tests/unit/test_cli.py::test_list_dash_dash_owner[1/2] PASSED
tests/unit/test_cli.py::test_list_dash_dash_owner[2/2] PASSED
====================== 112 tests deselected ======================
=========== 12 passed, 112 deselected in 0.16 seconds ============
```

你可以重复一个测试子集，或者某一个测试，甚至可以让测试在晚上重复1000 次。同时可以设置成让测试遇错就停止。

pytest-xdist：并行运行测试
pytest-xdist: Run Tests in Parallel

通常测试都是依次执行的，因为有些资源一次只能被一个测试用例访问。如果你的测试不需要访问共享资源，那么就可以通过并行运行来提速。pytest-xdist 插件可以实现这个功能。你可以指定处理器进程数目来同时运行多个测试，甚至可以把测试放在多台机器上运行。

下面的例子至少需要 1 秒的时间来运行，使用参数后它将运行 10 次。

```
appendices/xdist/test_parallel.py
import pytest
import time

@pytest.mark.parametrize('x', list(range(10)))
def test_something(x):
    time.sleep(1)
```

执行完至少需要 10 秒钟。

```
$ pip install pytest-xdist
$ cd /path/to/code/appendices/xdist
$ pytest test_parallel.py
==================== test session starts =====================
plugins: xdist-1.20.0, forked-0.2
collected 10 items

test_parallel.py ..........
================= 10 passed in 10.07 seconds =================
```

pytest-xdist 插件的 -n numprocesses 选项可以指定运行测试的处理器进程数，-n auto 选项可以自动侦测系统里的 CPU 数目。以下是上例在多个处理器上运行的情况。

```
$ pytest -n auto test_parallel.py
==================== test session starts =====================
plugins: xdist-1.20.0, forked-0.2
gw0 [10] / gw1 [10] / gw2 [10] / gw3 [10]
scheduling tests via LoadScheduling
..........
================= 10 passed in 4.27 seconds ==================
```

我们不能指望有几个处理器就能快多少倍，因为总会有一些额外的开销。

但很多测试场景允许并行测试，如果测试时间较长，就可以大幅节省时间。

pytest-xdist 插件的功能远比我们介绍的多，比如把测试全部转移到另外的机器上运行等，更多的用法请阅读 PyPI 上的 pytest-xdist 文档[1]。

pytest-timeout：为测试设置时间限制
pytest-timeout: Put Time Limits on Your Tests

正常情况下，pytest 里的测试是没有时间限制的。如果测试中涉及会消失的资源，比如 Web 服务，那么最好为测试加上时间限制。

pytest-timeout 插件[2]就提供了这个功能。它允许你在命令行指定超时时间，或者直接在测试代码中标注超时时间。

测试用例上标注的超时时间优先级高于命令行上的超时时间优先级，所以测试时间可能长于或短于命令行的设置。

让我们看看之前的例子（设置 1 秒钟休眠）再加上半秒超时限制后的运行情况。

```
$ cd /path/to/code/appendices/xdist
$ pip install pytest-timeout
$ pytest --timeout=0.5 -x test_parallel.py
===================== test session starts =====================
plugins: xdist-1.20.0, timeout-1.2.0, forked-0.2
timeout: 0.5s method: signal
collected 10 items
test_parallel.py F
=========================== FAILURES ==========================
_____ test_something[0] _____
x = 0
    @pytest.mark.parametrize('x', list(range(10)))
    def test_something(x):
>       time.sleep(1)
E       Failed: Timeout >0.5s
test_parallel.py:6: Failed
!!!!!!!!!!!!! Interrupted: stopping after 1 failures !!!!!!!!!!!!!
==================== 1 failed in 0.68 seconds ====================
```

---

[1] https://pypi.python.org/pypi/pytest-xdist
[2] https://pypi.python.org/pypi/pytest-timeout

使用-x选项会使测试遇错即停止运行。

C.2 改善输出效果的插件
Plugins That Alter or Enhance Output

有些插件不改变测试运行方式，只改变测试结果的显示方式。

pytest-instafail：查看错误的详细信息
pytest-instafail: See Details of Failures and Errors

通常 pytest 会显示每个测试的运行状态，当所有测试运行完毕后，pytest 将显示错误和失败用例的堆栈信息。如果测试很快运行完，这可能不是问题；但如果测试需要很长时间，那么我们希望一出错就能看到堆栈回溯信息，而不是等到所有测试都运行完。pytest-instafail 插件[1]就提供了这样的功能。使用--instafail 选项后，测试的失败和错误信息会马上显示出来。

下面是一个常规的测试用例，测试失败信息显示在最后。

```
$ cd /path/to/code/appendices/xdist
$ pytest --timeout=0.5 --tb=line --maxfail=2 test_parallel.py
=================== test session starts ===================
plugins: xdist-1.20.0, timeout-1.2.0, forked-0.2
timeout: 0.5s method: signal
collected 10 items

test_parallel.py FF
========================= FAILURES =========================
/path/to/code/appendices/xdist/test_parallel.py:6: Failed: Timeout
>0.5s
/path/to/code/appendices/xdist/test_parallel.py:6: Failed: Timeout
>0.5s
!!!!!!!!!! Interrupted: stopping after 2 failures !!!!!!!!!!
================ 2 failed in 1.20 seconds =================
```

接下来是使用--instafail 选项的情况。

```
$ pytest --instafail --timeout=0.5 --tb=line --maxfail=2
test_parallel.py
```

---

[1] https://pypi.python.org/pypi/pytest-instafail

```
=================== test session starts ===================
plugins: xdist-1.20.0, timeout-1.2.0, instafail-0.3.0, forked-0.2
timeout: 0.5s method: signal
collected 10 items

test_parallel.py F

/path/to/code/appendices/xdist/test_parallel.py:6: Failed: Timeout
>0.5s

test_parallel.py F

/path/to/code/appendices/xdist/test_parallel.py:6: Failed: Timeout
>0.5s
!!!!!!!!!! Interrupted: stopping after 2 failures !!!!!!!!!!
================= 2 failed in 1.19 seconds =================
```

--instafail 选项对那些运行时间长且需要人监视输出结果的测试非常有效。它可以让我们立即阅读错误及堆栈回溯信息，而不必手动停止测试，以影响后续测试的运行。

pytest-sugar：显示色彩和进度条
pytest-sugar: Instafail + Colors + Progress Bar

pytest-sugar 插件可以输出彩色字符。它也能在运行过程中显示错误和失败用例的堆栈回溯信息，而且还能在屏幕右侧显示进度条。

图 C.1 中的测试没有使用 pytest-sugar 插件。

```
[(venv) $ cd /Users/okken/code/ch7/tasks_proj_v2/
[(venv) $ pytest
=========================== test session starts ===========================
plugins: xdist-1.20.0, mock-1.6.2, forked-0.2
collected 62 items

tests/func/test_add.py ...
tests/func/test_add_variety.py .............................
tests/func/test_add_variety2.py ...........
tests/func/test_api_exceptions.py .........
tests/func/test_unique_id.py .
tests/unit/test_cli.py .....
tests/unit/test_task.py ....

======================= 62 passed in 0.24 seconds =======================
```

图 C.1　未使用 pytest-sugar 插件的测试示例

图 C.2 中的测试使用了该插件。

```
[(venv) $ pytest
Test session starts (platform: darwin, Python 3.6.2, pytest 3.2.1, pytest-sugar 0.9.0)
rootdir: /Users/okken/code/ch7/tasks_proj_v2, inifile: tox.ini
plugins: xdist-1.20.0, sugar-0.9.0, mock-1.6.2, forked-0.2
========================= test session starts =========================
 tests/func/test_add.py ✓✓✓                                        5%
 tests/func/test_add_variety.py ✓✓✓✓✓✓✓✓✓✓✓✓✓✓✓✓✓✓✓✓✓✓✓✓✓✓✓✓      50%
 tests/func/test_add_variety2.py ✓✓✓✓✓✓✓✓✓✓                        69%
 tests/func/test_api_exceptions.py ✓✓✓✓✓✓✓                         84%
 tests/func/test_unique_id.py ✓                                    85%
 tests/unit/test_cli.py ✓✓✓✓✓                                      94%
 tests/unit/test_task.py ✓✓✓                                      100%

Results (0.50s):
       62 passed
```

图 C.3 使用了 pytest-sugar 插件的测试示例

使用 pytest-sugar 插件后，通过的测试用例后面会打钩，失败的测试用例后会出现叉号，而且屏幕右侧会实时显示进度条。

pytest-emoji：为测试增添一些乐趣
pytest-emoji: Add Some Fun to Your Tests

pytest-emoji 插件允许你把所有测试状态字符替换为表情符号。如果插件作者定义的表情符不能让你满意，还可以自己来设置表情符号。这个插件的技术含量并不高，但是很有趣。读者也可以根据它编写自己的插件。

下面这个例子包含了所有可能的测试结果。我们先看看关闭堆栈回溯信息的输出结果（见图 C.3）。

```
[(venv) $ cd /Users/okken/code/appendices/outcomes/
[(venv) $ pytest --tb=no
============================ test session starts ============================
collected 6 items

test_outcomes.py .FxXsE

==== 1 failed, 1 passed, 1 skipped, 1 xfailed, 1 xpassed, 1 error in 0.07 seconds ====
```

图 C.3 关闭堆栈回溯信息的输出结果

再来看看使用详细信息选项 -v 的结果（见图 C.4）。

附录 C 常用插件

```
[(venv) $ pytest --tb=no -v
========================= test session starts =========================
collected 6 items

test_outcomes.py::test_pass PASSED
test_outcomes.py::test_fail FAILED
test_outcomes.py::test_xfail xfail
test_outcomes.py::test_xpass XPASS
test_outcomes.py::test_skip SKIPPED
test_outcomes.py::test_error ERROR

==== 1 failed, 1 passed, 1 skipped, 1 xfailed, 1 xpassed, 1 error in 0.06 seconds =====
```

图 C.4 使用详细信息选项 -v 的结果

现在，让我们使用表情符号选项 --emoji（见图 C.5）。

```
[(venv) $ pytest --tb=no --emoji
========================= test session starts =========================
plugins: emoji-0.1.0
collected 6 items

test_outcomes.py 😀😤😑😑😴😖

==== 1 failed, 1 passed, 1 skipped, 1 xfailed, 1 xpassed, 1 error in 0.07 seconds ====
```

图 C.5 使用表情符号选项 --emoji 的结果

下面是同时使用表情符号选项 --emoji 和详细信息选项 -v 的结果（见图 C.6）。

```
[(venv) $ pytest --tb=no -v --emoji
========================= test session starts =========================
plugins: emoji-0.1.0
collected 6 items

test_outcomes.py::test_pass PASSED 😀
test_outcomes.py::test_fail FAILED 😤
test_outcomes.py::test_xfail xfail 😑
test_outcomes.py::test_xpass XPASS 😑
test_outcomes.py::test_skip SKIPPED 😴
test_outcomes.py::test_error ERROR 😖

==== 1 failed, 1 passed, 1 skipped, 1 xfailed, 1 xpassed, 1 error in 0.06 seconds =====
```

图 C.6 同时使用 --emoji 和 -v 的结果

这是一个简单有趣的插件，可别小瞧它。它通过 hook 函数来改变表情符号，是少数几个展示了如何在插件中添加 hook 函数的 pytest 插件。

pytest-html：为测试生成 HTML 报告
pytest-html: Generate HTML Reports for Test Sessions

pytest-html 插件[1]对于持续集成或长时间运行的测试非常有用。它可以为 pytest 测试生成一个显示测试结果的网页。这个 HTML 报告可以对测试结果（通过、跳过、失败、错误、预期失败、预期失败但通过）进行筛选，还可以按测试名称、持续时间、结果状态来排序。HTML 报告还可以定制一些元素，如数据集的截图等。如果你要制作测试报告，不妨试一试 pytest-html。

pytest-html 插件的用法很简单，你只需加上 --html=report_name.html 选项即可。

```
$ cd /path/to/code/appendices/outcomes
$ pytest --html=report.html
====================== test session starts ======================
metadata: ...
collected 6 items

test_outcomes.py .FxXsE
 generated html file: /path/to/code/appendices/outcomes/report.html
============================ ERRORS =============================
_____ ERROR at setup of test_error _____
    @pytest.fixture()
    def flaky_fixture():
>       assert 1 == 2
E       assert 1 == 2
test_outcomes.py:24: AssertionError
=========================== FAILURES ============================
_____ test_fail _____
    def test_fail():
>       assert 1 == 2
E       assert 1 == 2
test_outcomes.py:8: AssertionError
1 failed, 1 passed, 1 skipped, 1 xfailed, 1 xpassed, 1 error in 0.08 seconds
$ open report.html
```

这将生成一个报告，包含测试会话信息、测试结果和汇总信息。

图 C.7 展示了测试会话信息和汇总信息。

---

[1] https://pypi.python.org/pypi/pytest-html

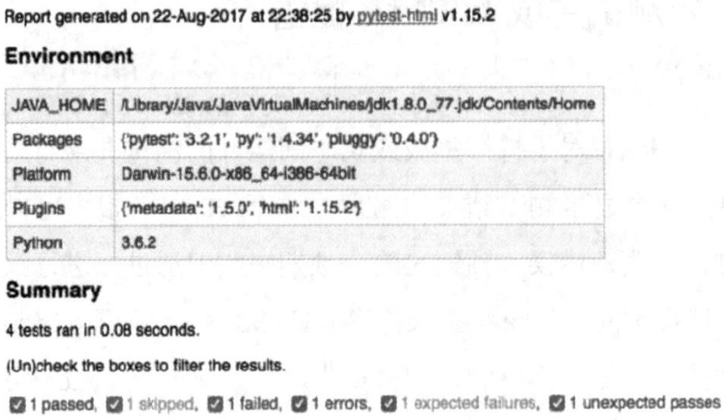

Report generated on 22-Aug-2017 at 22:38:25 by pytest-html v1.15.2

Environment

JAVA_HOME	/Library/Java/JavaVirtualMachines/jdk1.8.0_77.jdk/Contents/Home
Packages	{'pytest': '3.2.1', 'py': '1.4.34', 'pluggy': '0.4.0'}
Platform	Darwin-15.6.0-x86_64-i386-64bit
Plugins	{'metadata': '1.5.0', 'html': '1.15.2'}
Python	3.6.2

Summary

4 tests ran in 0.08 seconds.

(Un)check the boxes to filter the results.

☑ 1 passed, ☑ 1 skipped, ☑ 1 failed, ☑ 1 errors, ☑ 1 expected failures, ☑ 1 unexpected passes

图 C.7　示例的测试会话信息和汇总信息

图 C.8 展示了测试结果和汇总信息。

(Un)check the boxes to filter the results.

☑ 1 passed, ☑ 1 skipped, ☑ 1 failed, ☑ 1 errors, ☑ 1 expected failures, ☑ 1 unexpected passes

Results

Show all details / Hide all details

▲ Result	Test
Error (hide details)	test_outcomes.py::test_error::setup

```
@pytest.fixture()
    def flaky_fixture():
>       assert 1 == 2
E       assert 1 == 2

test_outcomes.py:29: AssertionError
```

Failed (hide details)	test_outcomes.py::test_fail

```
def test_fail():
>       assert 1 == 2
E       assert 1 == 2

test_outcomes.py:9: AssertionError
```

XFailed (hide details)	test_outcomes.py::test_xfail

```
@pytest.mark.xfail()
    def test_xfail():
>       assert 1 == 2
E       assert 1 == 2

test_outcomes.py:14: AssertionError
```

XPassed (hide details)	test_outcomes.py::test_xpass

```
No log output captured.
```

Skipped (hide details)	test_outcomes.py::test_skip::setup

```
('test_outcomes.py', 21, 'Skipped: unconditional skip')
```

Passed (show details)	test_outcomes.py::test_pass

图 C.8　示例的测试结果和汇总信息

这个报告包含 JavaScript 代码，可以对测试结果过滤和排序，你也可以在报告上添加额外的信息，包括图片。如果有上述需求，这个插件值得一试。

C.3　静态分析用的插件
Plugins for Static Analysis

静态分析工具可以在不运行代码的情况下做检查。Python 社区已经在使用这类工具。这里介绍几个插件，允许使用静态分析工具同时检查测试覆盖的代码和测试用例本身。静态分析失败同样会被显示为测试失败。

pytest-pycodestyle 和 pytest-pep8：Python 代码风格检查
pytest-pycodestyle, pytest-pep8: Python's Style Guide

PEP 8 是 Python 的代码风格指南[1]。Python 标准库代码要求遵循它的规范，大部分的 Python 开发者和开源项目也都遵循它的规范。pycodestyle[2] 命令行工具可以用来检查 Python 代码是否遵循 PEP 8 规范。安装 `pytest-pycodestyle` 插件[3]后，在命令行使用 `--pep8` 选项，pytest 将会调用 `pycodestyle` 检查测试代码和测试覆盖代码是否符合 PEP 8 规范。pycodestyle 工具更名之前叫 pep8[4]，所以，实际上 `pytest-pep8`[5]就是 `pytest-pycodestyle` 的老版本，有兴趣的读者也可以尝试。

pytest-flake8：更多的风格检查
pytest-flake8: Check for Style Plus Linting

`pep8` 只检查代码风格，而 `flake8` 会做更多静态分析检查。flake8[6]集成了

---

[1] https://www.python.org/dev/peps/pep-0008
[2] https://pypi.python.org/pypi/pycodestyle
[3] https://pypi.python.org/pypi/pytest-pycodestyle
[4] https://pypi.python.org/pypi/pep8
[5] https://pypi.python.org/pypi/pytest-pep8
[6] https://pypi.python.org/pypi/flake8

多种 Python 代码风格、多种静态分析工具的检查标准。它可以定制许多选项。通过 pytest-flake8 插件[1]，可以调用 flake8 帮助分析源码，包括测试源码，如果检查出异常，则会引起测试失败。它会检查 PEP 8，也会检查逻辑错误。在命令行中指定 flake8 选项后，就可以在 pytest 测试会话里运行 flake8 了。你还可以扩展 flake8 来添加更多的检查，比如 flake8-doctrings[2]，它可以针对 PEP 257[3] 检查 Python 代码中的文档字符串是否符合规范。

C.4 Web 开发用的插件
Plugins for Web Development

Web 项目有特定的测试逻辑。pytest 也不能把测试 Web 应用变得简单，但有几个插件可以帮上忙。

pytest-selenium：借助浏览器完成自动化测试
pytest-selenium: Test with a Web Browser

selenium 是一个可以自动运行 web 浏览器的工具。pytest-selenium 插件[4]是 Python 版本的 selenium。借助它可以启动一个 web 浏览器，打开网址 URL，运行 web 应用，填充表单，等等。你也可以通过编写测试代码来控制浏览器测试网站和 web 应用。

pytest-django：测试 Django 应用
pytest-django: Test Django Applications

Django 是很流行的基于 Python 的 web 开发框架。它自身包含用于测试的 hook 函数，允许你测试 Django 应用的各个组件，而不需要使用其他基于浏览

---

[1] https://pypi.python.org/pypi/pytest-flake8
[2] https://pypi.python.org/pypi/flake8-docstrings
[3] https://www.python.org/dev/peps/pep-0257
[4] https://pypi.python.org/pypi/pytest-selenium

器的测试。默认情况下，Django 测试使用的是 unittest。`pytest-django` 插件[1]可以让你使用 pytest 的全部功能。这个插件也包含一些帮助方法和 fixture，可以提高测试代码编写的速度。

pytest-flask：测试 Flask 应用
pytest-flask: Test Flask Applications

Flask 是另一个流行的 Web 框架（有时也称微框架）。`pytest-flask` 插件[2]提供了一组很有用的 fixture 来帮助测试 Flask 应用。

---

[1] https://pypi.python.org/pypi/pytest-django
[2] https://pypi.python.org/pypi/pytest-flask

附录 D

打包和发布 Python 项目
Packaging and Distributing Python Projects

项目的打包和发布很重要。大部分 Python 开发者对这一块并不熟悉，实际上，我们需要严肃地看待这个问题。毕竟，共享代码也是 Python 开发工作的一部分。因此，合理地使用 Python 内置的工具来共享代码很重要。虽然这是一个很大的话题，但由于受篇幅的限制，所以我无法全面介绍。这里只介绍常规的共享代码的方法。掌握这些方法后，至少你不必再用电子邮件发送压缩的文件和模块了。

我将介绍如何设置项目让它可以通过 `pip` 安装；如何以源码形式发布项目；如何将项目打包成 wheel 文件。这些技巧足以让你在小型团队内部共享代码。如果你还希望通过 PyPI 将代码共享到互联网上，请阅读我推荐的文档。现在让我们开始吧。

D.1 创建可安装的模块
Creating an Installable Module

先学习如何让一个小项目可以被 `pip` 安装。我以一个只有单一模块的项目为例。实际项目通常不会这么简单,我选择这个示例只是为了展示如何创建一个可维护的项目,以及 setup.py 文件可以多么简单。下面是一个简单的目录结构:

```
some_module_proj/
├── setup.py
└── some_module.py
```

我们想要共享的代码在 some_module.py 文件里:

appendices/packaging/some_module_proj/some_module.py
```python
def some_func():
    return 42
```

要使其能被 pip 安装,我们需要一个 setup.py 文件。下面是最简洁的 setup.py 代码:

appendices/packaging/some_module_proj/setup.py
```python
from setuptools import setup

setup(
    name='some_module',
    py_modules=['some_module']
)
```

一个目录、一个模块、一个 setup.py 文件,就足以使项目能够被 pip 安装了。

```
$ cd /path/to/code/appendices/packaging
$ pip install ./some_module_proj
Processing ./some_module_proj
Installing collected packages: some-module
  Running setup.py install for some-module ... done
Successfully installed some-module-0.0.0
```

现在可以在 Python 程序里(或者从测试用例里)使用 some_module 了。

```
$ python
Python 3.6.1 (v3.6.1:69c0db5050, Mar 21 2017, 01:21:04)
[GCC 4.2.1 (Apple Inc. build 5666) (dot 3)] on darwin
Type "help", "copyright", "credits" or "license" for more information.
```

```
>>>from some_module import some_func
>>>some_func()
42
>>>exit()
```

这是一种理想化的情况。在实际工作中，更常见的情况是要将项目进行打包。下一节将介绍如何修改 setup.py 文件来完成项目打包。

D.2 创建可安装的包
Creating an Installable Package

先创建一个以包名命名的目录，然后将__init__.py 文件和相关模块一同放入该目录里。

```
$ tree some_package_proj/
some_package_proj/
├──setup.py
└──src
    └──some_package
        ├──__init__.py
        └──some_module.py
```

some_module.py 文件的内容不变。__init__.py 文件需要把模块的功能通过包命名空间暴露给外部。有很多方式可以做到这一点，请阅读 Python 文档中关于该主题的两节内容[1]。

如果在__init__.py 文件里这样写：

```
import some_package.some_module
```

那么调用端的代码就必须指明 some_module。

```
import some_package
some_package.some_module.some_func()
```

但是我认为 some_module.py 是 API 功能的一部分，应该仅提供包这一层级

---

[1] https://docs.python.org/3/tutorial/modules.html#packages

的信息。所以，我们应该这样写：

```
appendices/packaging/some_package_proj/src/some_package/__init__.py
from some_package.some_module import *
```

现在调用端代码可以这样写了：

```
import some_package
some_package.some_func()
```

还需要对 setup.py 文件稍做修改。

```
appendices/packaging/some_package_proj/setup.py
from setuptools import setup, find_packages

setup(
    name='some_package',
    packages=find_packages(where='src'),
    package_dir={'': 'src'},
)
```

以后再调用就不需要提到 py_modules 了，只需指明包。现在它可以被安装了。

```
$ cd /path/to/code/appendices/packaging
$ pip install ./some_package_proj/
Processing ./some_package_proj
Installing collected packages: some-package
  Running setup.py install for some-package ... done
Successfully installed some-package-0.0.0
```

而且可以直接使用。

```
$ python
Python 3.6.1 (v3.6.1:69c0db5050, Mar 21 2017, 01:21:04)
[GCC 4.2.1 (Apple Inc. build 5666) (dot 3)] on darwin
Type "help", "copyright", "credits" or "license" for more information.
>>> from some_package import some_func
>>> some_func()
42
```

项目可以被安装了，而且很容易进行模块调用。你可以在与 src 同级的目录中添加 tests 目录并且放入测试用例。但 setup.py 文件还缺少重要内容，它还不能正常发布源码或创建 wheel 文件。不过，我们只需要略做修改即可。

D.3　创建源码发布包和 Wheel 文件
Creating a Source Distribution and Wheel

如果只是个人使用，上一节介绍的方法足以创建源码发布包和 wheel 文件了。下面来试试。

```
$ cd /path/to/code/appendices/packaging/some_package_proj/
$ python setup.py sdist bdist_wheel
running sdist
...
warning: sdist: standard file not found:
    should have one of README, README.rst, README.txt

running check
warning: check: missing required meta-data: url

warning: check: missing meta-data:
    either (author and author_email)
    or (maintainer and maintainer_email) must be supplied
running bdist_wheel
...
$ ls dist
some_package-0.0.0-py3-none-any.whl some_package-0.0.0.tar.gz
```

虽然出现了一些警告信息，但我们还是成功创建了 .whl 文件和 .tar.gz 文件。下面尝试消除这些警告。

步骤如下。

- 添加 README、README.rst 或者 README.txt 文件。
- 补充配置中的 url 数据。
- 补充作者和作者邮箱（或者维护者和维护者邮箱）信息。

同时还要添加以下信息。

- 版本号。
- 软件使用许可。
- 变更记录。

README 文件告诉用户如何使用你的包。url、作者信息告诉用户遇到问题时与谁联系。软件使用许可告诉用户使用包的注意事项（分发、贡献、复用代码有哪些限制）。如果你不希望开源，应当在软件使用许可中写清楚限制。如果开源，则推荐访问 https://choosealicense.com 选择合适的软件许可。

添加这些信息用不了多长时间，下面是一个精简的例子。

setup.py 文件：

appendices/packaging/some_package_proj_v2/setup.py
```python
from setuptools import setup, find_packages

setup(
    name='some_package',
    description='Demonstrate packaging and distribution',

    version='1.0',
    author='Brian Okken',
    author_email='brian@pythontesting.net',

url='https://pragprog.com/book/bopytest/python-testing-with-pytest',

    packages=find_packages(where='src'),
    package_dir={'': 'src'},
)
```

许可条文应该放到 LICENSE 文件里。本书的所有代码都使用如下许可：

appendices/packaging/some_package_proj_v2/LICENSE
```
Copyright (c) 2017 The Pragmatic Programmers, LLC

All rights reserved.

Copyrights apply to this source code.

You may use the source code in your own projects, however the source code
may not be used to create commercial training material, courses, books,
articles, and the like. We make no guarantees that this source code is
fit
for any purpose.
```

README.rst 文件的内容如下：

appendices/packaging/some_package_proj_v2/README.rst
```
====================================================
some_package: Demonstrate packaging and distribution
```

```
==================================================
``some_package`` is the Python package to demostrate how easy it is
to create installable, maintainable, shareable packages and
distributions.
It does contain one function, called ``some_func()``.

.. code-block

>>> import some_package
>>> some_package.some_func()
42

That's it, really.
```

README.rst 使用 reStrucuredText[1]软件排版。我习惯从现成的开源项目中复制 README.rst 文件，删掉那些我不需要的条文，然后根据自己项目的要求进行修改。

如果是一个严肃的项目，建议逐字输入 README。个人项目复制粘贴修改就可。

再添加一个记录项目变更的日志文件。下面是一个例子：

appendices/packaging/some_package_proj_v2/CHANGELOG.rst
```
Changelog
=========
--------------------------------------------------------
1.0
---
Changes:
~~~~~~~~
- Initial version.
```

请访问 http://keepachangelog.com，阅读编写变更日志的建议。本书 tasks_proj 的所有变更记录都放在 CHANGELOG.rst 文件里。

让我们看看这些改动是否足以消除警告。

```
$ cd /path/to/code/appendices/packaging/some_package_proj_v2
$ python setup.py sdist bdist_wheel
running sdist
```

---

[1] http://docutils.sourceforge.net/rst.html

```
running build
running build_py
creating build
creating build/lib
creating build/lib/some_package
copying src/some_package/__init__.py
  ->build/lib/some_package
copying src/some_package/some_module.py
  ->build/lib/some_package
installing to build/bdist.macosx-10.6-intel/wheel
running install
running install_lib
creating build/bdist.macosx-10.6-intel
creating build/bdist.macosx-10.6-intel/wheel
creating build/bdist.macosx-10.6-intel/wheel/some_package
copying build/lib/some_package/__init__.py
  ->build/bdist.macosx-10.6-intel/wheel/some_package
copying build/lib/some_package/some_module.py
  ->build/bdist.macosx-10.6-intel/wheel/some_package
running install_egg_info
Copying src/some_package.egg-info to
build/bdist.macosx-10.6-intel/wheel/some_package-1.0-py3.6.egg-info
running install_scripts
creating
build/bdist.macosx-10.6-intel/wheel/some_package-1.0.dist-info/WHEEL

$ ls dist
some_package-1.0-py3-none-any.whl some_package-1.0.tar.gz
```

很好，没有警告了。

现在可以把 .whl 文件和 .tar.gz 文件放到同一个目录，然后运行 `pip install`。

```
$ cd /path/to/code/appendices/packaging/some_package_proj_v2
$ mkdir ~/packages/
$ cp dist/some_package-1.0-py3-none-any.whl ~/packages
$ cp dist/some_package-1.0.tar.gz ~/packages
$ pip install --no-index --find-links=~/packages some_package
Collecting some_package
Installing collected packages: some-package
Successfully installed some-package-1.0
$ pip install --no-index --find-links=./dist some_package==1.0
Requirement already satisfied: some_package==1.0 in
  /path/to/venv/lib/python3.6/site-packages
$
```

现在可以为本地项目创建自己的包，它安装起来就像从 PyPI 安装那样容易。

D.4 创建可以从 PyPI 安装的包
Creating a PyPI-Installable Package

如果希望在 PyPI 上发布自己的包，那么需要在 setup.py 中添加更多的配置。同时还需要类似 Twine[1]这样的工具把包推送到 PyPI。Twine 是一组工具，它可让你更容易、更安全地与 PyPI 交互。它通过 HTTPS 认证保护 PyPI 密钥信息，并负责将包上传到 PyPI。

这些内容已经超出了本书的范围，细节请查阅 Python Packaging User Guide[2]（Python 程序用户分发指南）和 Python 文档中关于 PyPI 的部分[3]。

---

[1] https://pypi.python.org/pypi/twine
[2] https://python-packaging-user-guide.readthedocs.io
[3] https://docs.python.org/3/distutils/packageindex.html

附录 E

xUnit Fixture
xUnit Fixture

除了第 3 章介绍的 fixture，pytest 还支持 xUnit 风格的 fixture，类似于 Java 的 jUnit、C++ 的 cppUnit，以及其他语言的单元测试框架。

通常，xUnit 框架的测试流程是这样的：

```
setup()
test_function()
teardown()
```

每个测试用的运行都使用这个流程。pytest 逻辑上也支持这一类 fixture。pytest 也可以使用 setup() 和 teardown() 函数，只不过有些限制。

E.1　xUnit Fixture 的语法
Syntax of xUnit Fixtures

xUnit fixture 包括针对模块、函数、类等各个级别的 setup()/teardown() 方法。

setup_module()/teardown_module()

这些方法只在测试模块开始和结束时各运行一次。模块的参数是可选的。

setup_function()/teardown_function()

这些方法在核心测试函数之前和之后执行,不包含那些测试类方法。它们运行多次,每个测试函数都要执行一次。函数的参数是可选的。

setup_class()/teardown_class()

这些方法在每个测试类之前和之后执行。它们只运行一次。类的参数是可选的。

setup_method()/teardown_method()

这些方法在每个测试类的测试方法之前和之后执行。它们运行多次,每个测试方法只执行一次。方法的参数是可选的。

下面是 xUnit fixture 的一个例子。

```
appendices/xunit/test_xUnit_fixtures.py
def setup_module(module):
    print(f'\nsetup_module() for {module.__name__}')

def teardown_module(module):
    print(f'teardown_module() for {module.__name__}')

def setup_function(function):
    print(f'setup_function() for {function.__name__}')

def teardown_function(function):
    print(f'teardown_function() for {function.__name__}')

def test_1():
    print('test_1()')

def test_2():
    print('test_2()')

class TestClass:
    @classmethod
    def setup_class(cls):
        print(f'setup_class() for class {cls.__name__}')

    @classmethod
```

```python
    def teardown_class(cls):
        print(f'teardown_class() for {cls.__name__}')

    def setup_method(self, method):
        print(f'setup_method() for {method.__name__}')

    def teardown_method(self, method):
        print(f'teardown_method() for {method.__name__}')

    def test_3(self):
        print('test_3()')

    def test_4(self):
        print('test_4()')
```

我在 fixture 函数的参数中传入了模块、函数、类、方法的名字，以便在打印语句中输出。你不一定要用 module、function、cls、method 这样的参数名字，但是习惯上都这么用。

下面是一个测试会话，帮助你将流程可视化。

```
$ cd /path/to/code/appendices/xunit
$ pytest -s test_xUnit_fixtures.py
============ test session starts =============
plugins: mock-1.6.0, cov-2.5.1
collected 4 items

test_xUnit_fixtures.py
setup_module() for test_xUnit_fixtures
setup_function() for test_1
test_1()
.teardown_function() for test_1
setup_function() for test_2
test_2()
.teardown_function() for test_2
setup_class() for class TestClass
setup_method() for test_3
test_3()
.teardown_method() for test_3
setup_method() for test_4
test_4()
.teardown_method() for test_4
teardown_class() for TestClass
teardown_module() for test_xUnit_fixtures
========== 4 passed in 0.01 seconds ==========
```

E.2 混合使用 pytest Fixture 和 xUnit Fixture
Mixing pytest Fixtures and xUnit Fixtures

以下是混合使用 pytest Fixture 和 xUnit Fixture 的例子：

```
appendices/xunit/test_mixed_fixtures.py
import pytest

def setup_module():
    print('\nsetup_module() - xUnit')

def teardown_module():
    print('teardown_module() - xUnit')

def setup_function():
    print('setup_function() - xUnit')

def teardown_function():
    print('teardown_function() - xUnit\n')

@pytest.fixture(scope='module')
def module_fixture():
    print('module_fixture() setup - pytest')
    yield
    print('module_fixture() teardown - pytest')

@pytest.fixture(scope='function')
def function_fixture():
    print('function_fixture() setup - pytest')
    yield
    print('function_fixture() teardown - pytest')

def test_1(module_fixture, function_fixture):
    print('test_1()')

def test_2(module_fixture, function_fixture):
    print('test_2()')
```

这样做是允许的，但最好不要这么用，它会引起歧义。请看下面的例子：

```
$ cd /path/to/code/appendices/xunit
$ pytest -s test_mixed_fixtures.py
============ test session starts =============
plugins: mock-1.6.0, cov-2.5.1
collected 2 items
```

```
test_mixed_fixtures.py
setup_module() - xUnit
setup_function() - xUnit
module_fixture() setup - pytest
function_fixture() setup - pytest
test_1()
.function_fixture() teardown - pytest
teardown_function() - xUnit

setup_function() - xUnit
function_fixture() setup - pytest
test_2()
.function_fixture() teardown - pytest
teardown_function() - xUnit

module_fixture() teardown - pytest
teardown_module() - xUnit
========== 2 passed in 0.01 seconds ==========
```

这个例子说明 xUnit fixture 函数里的 module、function、method 参数都是可选的。我在函数定义里并没有使用它们，但是一样可以正常运行。

E.3 xUnit Fixture 的限制
Limitations of xUnit Fixtures

下面是使用 xUnit fixture 的一些限制，所以我更喜欢 pytest fixture。

- xUnit fixture 无法使用 –setup-show 和 –setup-plan 选项。这看起来无关紧要，但是当你有很多 fixture 要调试的时候，这些选项就很重要了。
- xUnit fixture 的作用范围不包括会话级别。它的最大作用范围是模块级别。
- xUnit 很难指定某个测试用例需要用的 fixture。
- 嵌套最多只有 3 层：模块、类和方法。
- 唯一可以用来优化 fixture 效果的方法是创建包含通用 fixture 需求的模块和类，这样所有测试用例都将使用该 fixture。
- xUnit fixture 不支持参数。你可以使用参数化的测试用例，但是 xUnit 的 fixture 不能参数化。

索引

Index

SYMBOLS
--instafail, 166
. (dot syntax), 1, 8
:: syntax, 40

A
a (pdb module), 128
add(), example of parametrized testing, 42–48
addopts, 114–115, 140
always print this, 85
and
 combining markers, 32
 running tests by name, 41
API (application programming interface)
 CLI interactions, xii
 mocks, 133–139
 types and functions, 30
approx(), 117
argparse, 133
args (pdb module), 128
_asdict(), 5
assert rewriting, 28–30
assert statements
 assert rewriting, 28–30
 exercise, 20
 simplicity of, xii
 using, 27–30
assert_called_once_with(), 137
attributes, setting and deleting, 86, 88
author field, packaging and distribution, 104, 178

author_email field, packaging and distribution, 104, 178
autouse
 initializing database for Task Project, 34
 parametrized testing of Task Project, 43
 using for fixtures, 61

B
base directory, 73
build-name-setter, 142
builtin fixtures, *see* fixtures, builtin

C
C (class scope), 58
cache, 77–84
--cache-clear option, 77, 80
--cache-show option, 77, 80
caches
 caching test sessions, 77–84
 installing from, 161
 using multiple versions, 161
capsys, 84
--capture=fd option, 14
--capture=method option, 9, 13
--capture=no option, 85
--capture=sys option, 14
category (warning), 93
change logs, 24, 178, 180
CHANGELOG.rst file, 24, 180
chdir(path), 86, 89
check_duration(), 83, 93, 110

cheese preference example, 85–89
class scope, 56, 72, 74, 183
classes
 defined, 40
 fixtures scope, 56, 72, 74, 183
 naming, 7, 114, 119–120
 parametrized testing, 46
 specifying fixtures, 61
 test discovery rules, 119
 testing single, 40
 xUnit fixtures, 183
CLI (command-line interface)
 adding options with pytest_addoption, 75
 configuring, 115–120
 interactions through API, xii
 mocks, 133–139
Click, 133–139
CliRunner, 137
cls object, 152
Cobertura, 147
code, for this book, xvi
code coverage, 129–132, 153
--collect-only option, 9–10
color displays, 1, 167
command-line interface, *see* CLI (command-line interface)
comments, 52, 56
configuration, 113–123
 command-line options, 115–120
 exercises, 122
 files, 113

索引 ◀ 217

listing options, 19, 114
marking xfail tests as failed, 38
minimum required version, 117
options, 19, 77, 114
with pytest.ini, 113–120
with pytestconfig, 75–77
registering markers, 116
test discovery, 114, 117–120
in test output, 7
testing multiple with tox, 139–142

conftest.py file
about, 113
additional configuration options, 115
in file structure, 25
as plugin, 50
sharing fixtures with, 50

context manager, mocks, 136

continuous integration
HTML report plugin, 169–171
installing plugins from local directory, 97
with Jenkins, 142–148
Python advantages, xi
specifying test directories, 119

cookiecutter-pytest-plugin, 110
--count option, 164
--cov-report option, 130
--cov=src option, 130
coverage.py, 129–132, 153
current directory, 83

D

d (pdb module), 128
data
passing fixture data to unittest functions, 152
test data with fixtures, 53–55
databases
initializing, 33
mocks, 135–139
parametrized fixtures, 67–69
running legacy unittest tests, 149
setup, 51, 67
db_type parameter, 68
debugging
display options, 125

with pdb module, 125–128, 153
stopping tests for, 12, 14
delattr(), 86
deleting
attributes, 86, 88
environment variables, 86–88
delitem(), 86
dictionaries
returning, 5
setting and deleting entries, 86, 88
directories
avoiding filename collisions, 113, 120
base, 73
cache, 80
changing current working, 86, 89
code coverage, 130
current, 83, 86, 89
dist, 109
distributing plugins from shared, 109
exercises, 48
help options, 20
installing plugins locally, 97
Jenkins testing, 143–148
packages file structure, 24
packaging plugins, 102
prepending paths, 89
rootdir, 7
specifying, 4–5, 39, 48, 73, 119
temporary, 34, 51, 55–61, 64, 71–75, 106
test discovery configuration options, 114, 117–120
test discovery rules, 119
in test output, 7
virtual environments, 156
dist directory, 109
distribution, 175–182
metadata, 104, 178
plugins, 109
resources, 102, 110, 182
setup.cfg file, 114
source, 109, 178–181
Django, 173
docstrings, 172

doctest
configuration options, 114
doctest_namespace fixture, 89–92
doctest_namespace, 89–92
doctest_optionflags, 114
dot syntax, 1, 8
down (pdb module), 128
duration
exercises, 93, 110
measuring with cache, 80–84
ordering test results by, 9, 18
duration_cache, 80–84
--durations=N option, 9, 18

E

E (errors)
defined, 8
displays, 8, 29
emojis, 168–169
entry_points field, packaging plugins, 104
environment variables
listing, 19
setting and deleting, 86–88
environments, see virtual environments
equivalent(), 43
errors, see also failing tests
defined, 8
displays, 8, 29, 166–171
emojis, 168–169
raising with monkeypatch fixture, 86
testing for expected, 30
traceback as they happen, 166
ExceptionInfo, 31
as excinfo, 30
exercises
configuration, 122
files and directories, 48
fixtures, 69, 93
installation, 20, 48
Jenkins, 153
markers, 48
mocks, 153
pdb module, 153
plugins, 110, 122
tox, 153
virtual environments, 20

218 ▶ 索引

--exit first option
    exercises, 153
    timeouts, 166
    using, 9, 12, 125
EXPRESSION option, 9–10
expressions, running tests with, 9–10, 41

F

F (failures)
    changing indicator plugin example, 98–110
    defined, 8
    displays, 8
F (function scope), 53, 58
failing tests, see also xfail tests
    assert rewriting, 28–30
    changing indicator plugin example, 98–110
    debugging with pdb module, 125–128
    defined, 8
    disallowing xpass, 114, 120
    displays, 1, 8, 29, 140, 166–171
    emojis, 168–169
    marking expected failures, 8, 37, 48
    reporting failing line, 15
    running first failed, 9, 14, 77–80
    running last failed, 9, 14, 77–80, 82, 125, 153
    stopping tests at, 9, 12, 125, 164
    stopping with --maxfail, 9, 13
    testing for expected exceptions, 30
    timeouts, 166
    tox, 140
    traceback as they happen, 166
    unittest tests and first failures, 153
fakes, see mocks
--ff option
    with cache, 77–80
    using, 9, 14
file descriptors, capture options, 14
filename (warning), 93
filename collisions, avoiding, 93, 113, 120
files
    avoiding filename collisions, 93, 113, 120
    downloading multiple with pip, 161
    exercises, 48
    file descriptors capture options, 14
    fixtures in individual, 50
    help options, 20
    naming conventions, 5, 7, 81, 114, 120
    packages file structure, 24
    specifying, 4–5, 40, 48
    specifying one, 7, 40, 48
    specifying only one test, 8
    test discovery rules, 114, 120
--find-links myplugins, 109
find-links=./some_plugins/, 97
--first-failed option
    with cache, 77–80
    using, 9, 14
fixture(), 49
fixtures, 49–70, see also fixtures, builtin
    autouse, 34, 43, 61
    changing scope, 59–61
    configuration options, 114
    defined, 25, 49
    directories, 25
    exercises, 69, 93
    help options, 19
    listing, 63
    marking with, 61
    mixing xUnit with pytest, 185
    multiple, 55
    names, 49, 63–64
    parametrized, 64–69, 153, 187
    passing data to unittest functions, 152
    as plugins, 95
    renaming, 63–64
    scope, 53, 56–61, 69, 72, 151
    session scope, 151
    setup and teardown with, 51–53, 58
    sharing with conftest.py file, 50
    specifying with usefixtures, 61
    as term, 50
    for test data, 53–55
    testdir, 105–108
    testing plugins, 105–108
    tracing execution with --setup-show, 52, 58, 69, 187
    xUnit, 183–187
--fixtures option, 63
fixtures, builtin, 71–93
    advantages, 71
    cache, 77–84
    capsys, 84
    doctest_namespace, 89–92
    exercises, 93
    monkeypatch, 85–89
    options, 76
    pytestconfig, 75–77
    recwarn, 92
    request, 65, 68, 81, 152
    scope, 72
    tmpdir, 34, 51, 55–61, 64, 71–75
    tmpdir_factory, 51, 55–61, 64, 71–75
flake8, 172
Flask, 173
floating point numbers, 117
fnmatch_lines, 106
--foo <value> option, 75
function scope
    cache, 82
    display, 53
    fixtures, 53, 56, 72
    tmpdir, 72
    xUnit fixtures, 183
functional tests
    defined, xiii
    directories, 24
functions, 23–48, see also fixtures; hook functions
    API calls and types, 30
    assert statements, 27–30
    importing, 25
    marking, 31–38
    names, 7, 114, 120
    parametrized testing, 42–48
    resources, 98
    test discovery rules, 114, 120
    testing for expected exceptions, 30
    testing single, 40
    xUnit fixtures, 183

G

getoption(), 122
GitHub, 82, 96–97
.gitignore file, 156

H

-h, using, 19
headers, test, 100
--help
    ini file options, 114
    listing options with, 9
    using, 19
    virtual environments, 157
hook functions
    adding to plugins, 169
    defined, 25
    directories, 25
    as plugins, 95
    pytest_addoption, 75, 122
    pytest_report_header(), 100
    pytest_report_teststatus(), 100
    resources, 98, 101
HTML
    code coverage reports, 130
    report plugin, 169–171
HTTPS, 182

I

id field
    checking, 43
    exercises, 48
    generating identifiers, 66
    optional, 45
    parametrized fixtures, 66
    parametrized testing, 43, 45, 47
importing
    docstring failures, 90–92
    functions, 25
    packages, 25–27
ini files and configuration, 7, 25, 113–123
__init__.py files
    about, 25, 113
    avoiding filename collisions, 120
    creating installable packages, 176
--instafail option, 166
installing
    from cache, 161
    coverage.py, 129
    exercises, 20, 48
    Jenkins plugins, 143
    MongoDB, 69, 126
    packages locally, 25–27, 97
    plugins and packages, 25–27, 96–97, 108–109, 159–161
    pytest, 3, 20
    from .tar.gz and .whl files, 96, 161
    tox, 141
    virtual environments, 155
integration tests, defined, xiii

J

Jenkins CI, 142–148, 153
--junit-prefix option, 145
--junit-xml option, 145
junit_suite_name, 145

K

-k option, 9–10, 41
key names, 81

L

l (pdb module), 128
-l option (print statements), 9, 14, 16, 79, 115, 125
lambda expression, 88
--last-failed option
    with cache, 77–80
    measuring duration with cache, 82
    using, 9, 14, 125, 153
--lf option
    with cache, 77–80
    measuring duration with cache, 82
    using, 9, 14, 125, 153
license field, packaging and distribution, 104, 178
LICENSE file, 24, 104, 178
licenses, 24, 104, 178
lineno (warning), 93
list (pdb module), 128
list begin, end (pdb module), 128
local variables
    printing output, 9, 14, 16, 79, 115, 125
    tox, 140

M

M (module scope), 58
-m option, using, 9, 11, 32
MagicMock, 137
maintainer field, packaging and distribution, 104, 178
maintainer_email field, packaging and distribution, 104, 178
makepyfile(), 106
MANIFEST.in file, 24
markers
    configuration options, 114–115
    exercises, 48
    expected failures, 8, 37, 48
    with fixtures, 61
    functions, 31–38
    help options, 19
    multiple, 12, 32
    names, 11
    registering, 116
    skipping tests with, 8, 34–37, 48
    --strict option, 115, 140
    timeouts, 165
    tox, 140
    unittest tests, 152
    using, 9, 11
--markers list, 116
markers option, 114, 140
MARKEXPR option, using, 9, 11
math example of doctest_namespace, 89–92
--maxfail option, 9, 13
members, accessing by name, 5
message (warning), 93
metadata
    HTML report plugin, 169
    packaging and distribution, 104, 178
method scope, 183
methods, *see also* fixtures
    method scope, 183
    names, 7, 114, 120
    parametrized testing, 46
    test discovery rules, 114, 120
    testing single, 41
    xUnit fixtures, 183
minversion, 114, 117
mocker, 136
mocks, 133–139, 153
module scope, 56, 72, 74, 183
modules
    creating installable, 175–176
    doctest_namespace fixture, 89–92

fixtures scope, 56, 72, 74, 183
packaging plugins, 104, 178
prepending paths, 89
specifying one, 40
test discovery rules, 119
xUnit fixtures, 183
MongoDB, 67–69, 126
monkeypatch, 85–89
--myopt option, 75

N

-n auto, 165
-n numprocesses, 165
name parameter, 63
namedtuple(), 3
names
　accessing members by, 5
　avoiding filename collisions, 113, 120
　doctest_namespace fixture, 89–92
　files, 5, 7, 81, 114, 120
　fixtures, 49, 63–64
　functions, 7, 114, 120
　key, 81
　markers, 11
　methods, 7, 114, 120
　naming conventions, 5, 7, 81, 119–120
　renaming fixtures, 63–64
　running tests by, 41
　test discovery rules, 119
__new__.__defaults__, 5
nice status indicators plugin example, 98–110
--no-index option (pip), 97, 109
nodeid, 81
nodes, 44
norecursedirs, 114, 118
normal print, usually captured, 85
not
　combining markers, 32
　running tests by name, 41
NUM, 73

O

O (opportunity), changing indicator plugin example, 98–110
ObjectID, 127

or
　combining markers, 32
　running tests by name, 41
output
　with capsys, 84
　options, 1, 8–9, 13, 85, 115, 117, 125
　plugins for enhancing, 166–171
　understanding, 1, 7–8

P

packages
　doctest_namespace fixture, 89–92
　file structure, 24
　importing functions, 25
　installing, 25–27, 97, 139, 159–161
　installing locally, 25–27, 97, 160
　installing with tox, 139
　listing, 160
　namespace, 177
　packaging tips, 102–105, 175–182
　resources, 110, 182
　setup.cfg file, 114
　specifying versions, 160
　from .tar.gz and .whl files, 96, 161
　uninstalling, 160
parallel tests, 164
param(), 47
parameters
　monkeypatch fixture, 86
　testing for expected exceptions, 30
parametrize(), 43–48
parametrized fixtures, 64–69, 153, 187
parametrized testing, 42–48, 187
passing tests, see also xpass tests
　defined, 8
　displays, 1, 8, 168–169
　dot syntax, 1, 8
　emojis, 168–169
paths
　configuration options, 114
　prepending, 86, 89
--pdb command, 125, 127
pdb module, 125–128, 153

PEP 257, 172
PEP 8, 172
pep8 command-line tool, 172
--pep8 option, 172
pip
　about, xv, 159–161
　distributing plugins from, 109
　downloading multiple files, 161
　ignoring PyPI option, 97, 109
　installing MongoDB, 126
　installing Task Project, 26
　installing plugins, 96–97
　installing pytest, 4
　packaging and distribution, 109, 175–182
　resources, 161
　tox and, 139
　uninstalling with, 160
　versions, 159
platform darwin, 7
pluggy, 7
plugins, 95–111
　changing test flow, 163–166
　configuration options, 115
　conftest.py as, 50
　creating, 98–110
　directories, 102
　exercises, 110, 122
　help options, 19
　hook functions, adding to, 169
　installing, 96–97, 108–109, 159–161
　installing from Git repository, 97
　installing from local directory, 97
　installing multiple versions, 97
　installing specific versions, 96
　Jenkins, 142
　listing, 160
　output enhancing, 166–171
　packaging and distribution, 102–105, 109, 175–182
　resources, 95–96
　static analysis, 171
　suggested, 163–173

testing, 101, 105–108, 122
testing manually, 101, 106, 122
uninstalling, 108, 160
versions, 96–97, 104
web development plugins, 172
pp expr (pdb module), 128
prepend parameter, 86
pretty printing, pdb module, 128
print expr (pdb module), 128
print statements
  with capsys, 84
  disabling, 85
  fixtures exercise, 69
  options, 9, 13, 16, 79, 115, 125
  pdb options, 128
progress bar, 167
py, 7
pycodestyle command-line tool, 172
pydocstyle, 172
pymongo, 126
PyPI, see Python Package Index (PyPI)
pytest, see also configuration; fixtures; functions; plugins
  advantages, xi, 1–2
  basics, 1–21
  installing, 3, 20
  minimum required version, 114, 117
  options, 9–20
  resources, 3, 98
  running, 4–9
  versions, xii, 7, 9, 19, 114, 117
pytest config object, 75–77
pytest-cov, 96, 129–132
pytest-django, 173
pytest-emojis, 168–169
pytest-flake8, 172
pytest-flask, 173
pytest-html, 169–171
pytest-instafail, 166
pytest-mock, 133–139
pytest-nice, 98–110
pytest-pep8, 172
pytest-pycodestyle, 172
pytest-repeat, 163

pytest-selenium, 172
pytest-sugar, 167
pytest-timeout, 165
pytest-xdist, 164
pytest.ini file, 25, 113–123
pytest11, 104
pytest_addoption, 75, 122
pytest_report_header(), 100
pytest_report_teststatus(), 100
pytestconfig, 75–77
pytester, 105–108, 122
Python
  about, xi
  PEP 8, 172
  resources, 102, 110, 128
  versions, xii, 4, 7, 133, 159
Python Package Index (PyPI)
  credentials, 182
  distribution on, 110, 182
  exercises, 110
  ignoring, 97, 109
  installing pytest, 3
  plugins from, 96, 165
  resources, 182
Python Packaging Authority, 157, 161
Python Packaging User Guide, 110, 182
python_classes, 114, 119–120
python_files, 114, 120
python_functions, 114, 120

Q
q (pdb module), 128
-q option, 9, 15
--quiet option, 9, 15
quit (pdb module), 128
quotes, parametrized testing, 44, 46

R
raising parameter, 86
random, 81
readability, xii, 45
README files, 24, 104, 178
recursion, configuration options, 114, 117–120
recwarn, 92
registering markers, 116
repeating tests plugin, 163
_replace(), 5

reports
  code coverage, 130
  HTML report plugin, 169–171
  Jenkins Test Results Analyzer, 142
request, 65, 68, 81, 152
requirements.txt file, 161
resources
  for this book, xvi, 154
  code coverage, 132
  docstrings, 172
  functions, 98, 101
  hook functions, 98, 101
  Jenkins, 147
  mocks, 137, 139
  packaging and distribution, 102, 110, 177, 182
  packaging namespace, 177
  pdb module, 128
  pip, 161
  plugins, 95–96
  pytest, 3, 98
  pytest-xdist, 165
  pytester, 122
  Python, 102, 110, 128
  tox, 142
  virtual environments, 157
ret, 106
rootdir, 7
-rs option, 37
-rsxX option, 115, 117, 140

S
S (session scope), 53, 58
s (skipped tests)
  defined, 8
  displays, 8, 36
-s option (print statements), 9, 13, 85
scope
  cache, 82
  changing, 59–61
  fixtures, 53, 56–61, 69, 72, 151, 183
  tmpdir, 59, 72
  tmpdir_factory, 59, 72, 74
  xUnit fixtures, 183, 187
scope parameter, 56
screenshots, HTML report plugin, 169
sdist, 109
Selenium, 172

session, *see also* session scope
   cache fixture, 77–84
   duration, 8, 80–84, 110
   HTML report plugin, 169–171
   repeating tests in, 163
   in test output, 7
session scope
   cache, 83
   changing, 59–61
   display, 53, 58
   fixtures, 53, 56, 72
   tmpdir_factory, 72
   unittest tests, 151
   xUnit fixtures, 187
setUp(), 152
setattr(), 86, 88
setenv(), 86, 88
setitem(), 86, 88
setting
   attributes, 86, 88
   environment variables, 86–88
setup, *see also* fixtures
   and --durations=N option, 19
   with fixtures, 51–53, 58
   packaging and distribution, 24, 103, 109, 114, 176–182
   passing ids with fixtures, 152
   --setup-show option, 52, 58, 69, 187
   with xUnit fixtures, 183–187
setup() (xUnit fixture), 183–187
setup() file (packaging plugins), 103
--setup-plan option, 187
--setup-show option, 52, 58, 69, 187
setup.cfg, 114
setup.py file
   distribution and setup.cfg file, 114
   installing local package, 160
   in package file structure, 24
   packaging and distribution, 103, 109, 114, 176–182
   tox, 139
setup_class(), 183
setup_function(), 183

setup_method(), 184
setup_module(), 183
--showlocals option, 9, 14, 16, 79, 125
skip(), 8, 34–36
skipif(), 8, 34, 36
skipping tests
   defined, 8
   displays, 8, 140
   emojis, 168–169
   exercises, 48
   with markers, 8, 34–37, 48
   tox, 140
smoke tests, 31–34
source distribution, 109, 178–181
speed
   exercise, 93
   measuring duration with cache, 80–84
   ordering tests by duration, 9, 18
   parallel tests, 164
spies, *see* mocks
src/ directory, 25
stack trace, *see* traceback
static analysis tools, 171
stderr
   with capsys, 84
   options, 9, 13, 85
stdout
   with capsys, 84
   options, 9, 13, 85
   testing plugins, 106
stopping
   tests with --maxfail, 9, 13
   tests with -x and --exit first, 9, 12, 125
   tests with pytest-repeat, 164
--strict option, 115, 140
strings
   parametrized testing, 44
   testing plugins, 106
stubs, *see* mocks
subcutaneous tests, defined, xiii
sugar plugin, 167
syspath_prepend(path), 86, 89
system tests, defined, xiii

T

t_after local variable, 17
t_before local variable, 17

t_expected local variable, 17
tar balls
   distribution from, 109, 178–181
   installing plugins and packages from, 96
.tar.gz
   distribution from, 109, 178–181
   installing plugins and packages from, 96
tasks
   installing, 25–27
   sample session, xii
Tasks project
   about, xii
   assert statements, 27–30
   builtin fixtures for, 71–93
   changing scope, 59–61
   changing status indicators plugin example, 98–110
   code coverage, 129–132, 153
   configuration, 113–123
   creating objects, 5
   debugging with pdb module, 125–128
   functions, 31–38
   installing locally, 25–27
   with Jenkins CI, 142–148
   legacy testing, 148–153
   marking expected failures, 37
   mocks, 133–139
   parametrized fixtures, 64–69
   parametrized testing, 42–48
   plugins, creating, 98–110
   plugins, using, 95–97, 163
   renaming fixtures, 63–64
   running subsets of tests, 38–42
   setup and teardown with fixtures, 51–53, 58
   setup options, 9–20
   skipping tests, 34–37
   smoke tests, 31–34
   source code, xvi
   structure, 3, 24
   test data with fixtures, 53–55
   test discovery configuration options, 118
   testing for expected exceptions, 30

testing multiple configurations with tox, 139–142
timer for, 61
--tb=auto option, 18
--tb=line option, 15, 17
--tb=long option, 18
--tb=native option, 18
--tb=no option, 12, 17
--tb=short option, 17, 115, 117, 140
--tb=style option, 9, 17, 125, 140
teardown, see also fixtures
  and --durations=N option, 19
  with fixtures, 51–53, 58
  with xUnit fixtures, 183–187
teardown() (xUnit fixture), 183–187
teardown_class(), 183
teardown_function(), 183
teardown_method(), 184
teardown_module(), 183
test classes, see classes
test discovery
  configuration options, 114, 117–120
  file structure, 25
  naming conventions, 5, 7
  rules, 119–120
test doubles, see mocks
test files, see files
test functions, see functions
test headers, 100
test methods, see methods
Test Results Analyzer, 142
test scope, 72
test_defaults(), 5
test_failing(), order and --first-failed option, 15
test_member_access(), 5
testdir, 105–108
testdir.runpytest(), 106
testing
  code coverage, 129–132, 153
  legacy, 148–153
  parallel tests, 164
  parametrized, 42–48, 187
  repeating tests per session, 163
  running subsets of tests, 38–42

smoke tests, 31–34
specifying files and directories, 4–5, 7–8, 10, 39–40
specifying only one test, 8
terms, xiii
testpaths, 114, 118
tests/ directory, 25
timeouts, 165
TinyDB, 51, 67
tmpdir
  defined, 64
  initializing database for Task Project, 34
  scope, 56–61, 72
  using, 51, 55, 71–75
tmpdir_factory, 51, 55–61, 64, 71–75
tox
  configuration, 113
  exercises, 153
  installing, 141
  installing plugins from local directory, 97
  Jenkins with, 147
  resources, 142
  specifying test directories, 119
  testing multiple configurations with, 139–142
tox.ini, 113, 139
traceback
  color and progress bar, 167
  displaying as failures happen, 166
  emojis plugin, 168–169
  fixtures, 53
  navigating in pdb module, 128
  options, 9, 12, 15, 17–18, 115, 117, 125
  tox, 140
  turning off, 12, 17
Twine, 182
types
  API calls and functions, 30
  database, 68
  parametrized testing, 44

U

u (pdb module), 128
Ubuntu and issues with venv, 155

uninstall something, 160
uninstalling
  packages, 160
  plugins, 108, 160
unit tests
  defined, xiii
  directories, 24
unittest
  Django support, 173
  mocks, 133–139
  running legacy tests, 148–153
up (pdb module), 128
url field, packaging and distribution, 104, 178
usefixtures, 61, 114

V

-v option
  changing indicator plugin example, 101
  checking for right tests, 11
  marking functions, 32
  parametrized tests and fixtures, 68
  running specific directories, classes, and tests, 39, 41
  skipping tests, 36
  using, 1, 8–9, 15, 125
  xfail tests, 38
venv, 4, 20, 155
--verbose option
  changing indicator plugin example, 101
  checking for right tests, 11
  marking functions, 32
  parametrized tests and fixtures, 68
  running specific directories, classes, and tests, 39, 41
  skipping tests, 36
  using, 1, 8–9, 15, 125
  xfail tests, 38
--version option, 9, 19
version field, packaging and distribution, 104, 178
versions
  checking, 160
  downloading multiple with pip, 161
  minimum pytest, 114, 117
  mock package, 133

pip, 159
plugins, 96–97, 104
pytest, xii, 7, 9, 19, 114, 117
Python, xii, 4, 7, 133, 159
specifying package, 160
in test output, 7
using multiple, 139–142, 159, 161
--version option, 9, 19
virtualenv, 155
virtual environments
advantages, 155
exercises, 20
installing, 155
installing plugins from local directory, 97
installing pytest, 4
installing tox, 141
Jenkins, 143–148
multiple, 139–142, 147, 161
pip and, 159
resources, 157
using, 155–157, 161
using multiple versions, 161
--version option, 9, 19
virtualenv
exercises, 20
installing pytest, 4
using, 155–157
versions, 155

W

warnings
with recwarn, 92
with warns(), 93
warns(), 93
web browser plugins, 172
web development plugins, 172
wheels
distributing from, 178–181
installing plugins and packages from, 96, 161
.whl
distributing from, 178–181
installing plugins and packages from, 96, 161
Windows
installing pytest, 4
platform in test output, 7
virtual environment set-up, 156

X

x (xfail)
defined, 8
displays, 8
X (xpass)
defined, 8
displays, 8

-x (–exit first) option
exercises, 153
timeouts, 166
using, 9, 12, 125
xUnit fixtures, 183–187
xdist plugin, 164
xfail tests
defined, 8
displays, 8, 140, 168–169
emojis, 168–169
exercises, 48
marking, 37
strict, 114, 120
tox, 140
xfail(), 8, 37
xfail_strict, 114, 120
XML, Jenkins, 142, 145
xpass tests
defined, 8
disallowing, 114, 120
displays, 8, 140, 168–169
emojis, 168–169
tox, 140

Y

yield, 34, 51

Z

zip files, installing plugins and packages from, 96, 108, 161